Asheville-Buncombe
Technical Community College
Learning Resources Center
340 Victoria Rd.
Asheville, NC 28801

RADIATION ENERGY TREATMENT OF WATER, WASTEWATER AND SLUDGE

A State-of-the-Art Report
by the Task Committee on Radiation Energy Treatment
Air and Radiation Management Committee
Environmental Engineering Division
American Society of Civil Engineers

Committee Members

Edward H. Bryan
Richard I. Dick
Jean F. Swinwood
Paul Kruger, Chairman

Dale A. Carlson
Gerry Hare
Tom Waite

DISCARDED

JUN 2 3 2025

Published by the
American Society of Civil Engineers
345 East 47th Street
New York, New York 10017-2398

ABSTRACT

Since ionizing radiation is one of the more technologically advanced forms of energy available to society, it has been applied to a great many activities such as: the sterilization of a wide range of consumer goods; preservation of food; chemical processing of complex materials; and medical diagnosis and therapy techniques. However, using radiation energy to ensure safe water supplies and municipal and industrial wastewater treatment is a relatively new civil engineering application. This committee report, *Radiation Energy Treatment of Water, Wastewater and Sludge*, is a state-of-the-art summary of need, progress, potential, and problems in the use of radiation energy as a treatment technology of water, wastewater and sludge. It discusses radiation sources with energy at or above ultraviolet light, and the appropriate application for each source type. In addition, the report identifies some of the areas where radiation energy can be used in combination with other water treatment techniques, examines some of the technical issues which would enhance its acceptance, and explores some of the economic constraints to wide-spread application.

Library of Congress Cataloging-in-Publication Data

Radiation energy treatment of water, wastewater, and sludge: a state-of-the-art report/by the Task Committee on Radiation Energy Treatment of Water and Wastewater, Air and Radiation Management Committee, Environmental Engineering Division, American Society of Civil Engineers.
  p. cm.
  Includes bibliographical references and index.
  ISBN 0-87262-901-5
  1. Water — Purification — Irradiation. 2. Sewage — Purification — Irradiation. 3. Sewage sludge — Irradiation.
  I. American Society of Civil Engineers. Task Committee on Radiation Energy Treatment of Water and Wastewater.
  TD476.R3 1992
  628.1'62—dc20   92-20609
                                                              CIP

The material presented in this publication has been prepared in accordance with generally recognized engineering principles and practices, and is for general information only. This information should not be used without first securing competent advice with respect to its suitability for any general or specific application.

The contents of this publication are not intended to be and should not be construed to be a standard of the American Society of Civil Engineers (ASCE) and are not intended for use as a reference in purchase specifications, contracts, regulations, statutes, or any other legal document.

No reference made in this publication to any specific method, product, process, or service constitutes or implies an endorsement, recommendation, or warranty thereof by ASCE.

ASCE makes no representation or warranty of any kind, whether express or implied, concerning the accuracy, completeness, suitability or utility of any information, apparatus, product, or process discussed in this publication, and assumes no liability therefor.

Anyone utilizing this information assumes all liability arising from such use, including but not limited to infringement of any patent or patents.

Authorization to photocopy material for internal or personal use under circumstances not falling within the fair use provisions of the Copyright Act is granted by ASCE to libraries and other users registered with the Copyright Clearance Center (CCC) Transactional Reporting Service, provided that the base fee of $1.00 per article plus $.15 per page is paid directly to CCC, 27 Congress Street, Salem, MA 01970. The identification for ASCE Books is 0-87262/92. $1 + .15. Requests for special permission or bulk copying should be addressed to Reprints/Permissions Department.

Copyright © 1992 by the American Society of Civil Engineers,
All Rights Reserved.
Library of Congress Catalog Card No: 92-20609
ISBN 0-87262-901-5
Manufactured in the United States of America.

TABLE OF CONTENTS

Section		Page
1.	INTRODUCTION	1
2.	FUNDAMENTALS OF RADIATION TREATMENT	2
3.	GAMMA RADIATION	7
4.	ELECTRON-BEAM RADIATION	13
5.	X-RAY APPLICATIONS	19
6.	ULTRAVIOLET RADIATION	21
7.	COMBINED PROCESSES	27
8.	APPLICATION ASSESSMENT	30
9.	ISSUES, CONCLUSIONS, and RECOMMENDATIONS	37
	REFERENCES	41

RADIATION ENERGY TREATMENT OF WATER, WASTEWATER AND SLUDGE
By the Task Committee on Radiation Energy Treatment
Air and Radiation Management Committee
Environmental Engineering Division
American Society of Civil Engineers

Section 1. INTRODUCTION

Ionizing radiation is one of the more technologically advanced forms of energy available to society. Applications of radiation energy have been under engineering development since the introduction of large-scale sources in the form of high-voltage machines and radioactive materials. Included among these applications are: the sterilization of a wide range of consumer goods and medical products; remote electricity supply as long-lived unattended batteries, non-destructive testing equipment, conservative tracers in almost any chemical, physical, and biological form; preservation of food; chemical processing of complex materials; and medical diagnosis and therapy techniques. The application of radiation energy for ensuring safe water supplies and municipal and industrial wastewater treatment is a relatively new addition to this list.

In the continuous development of methods for treatment of water, a diverse array of physical, chemical, and biological processes have been developed and put into practice as the technology was perfected and the economics allowed. In the quest for more efficient methods to purify water supplies and treat wastewater and sludge, great interest was shown in the potential for radiative energy following the initiation of the atomic era in the early 1900's. A large body of research experience was created in the subject of radiation chemistry and the prospects for civil engineering application to water treatment was investigated. Early evaluations in the 1950's included Dunn (1953), Lowe et al (1956), Ridenour and Armbruster (1956), and Narver (1957). A review of the advances in radiation treatment of wastes through 1975 was given by Feates and George (1975).

Radiation processing technology for industrial applications began in the early 1960's with the sterilization of medical products. It has since grown to an industry in which irradiators are employed in more than 46 countries to treat a variety of consumer and medical products, and to chemically modify a large number of materials. Its application as a waste treatment process is relatively new. As public concerns for the environment increase and regulatory requirements expand, radiation technology will play a more important role in the practice of civil and environmental engineering.

In 1989, the Task Committee on Radiation Energy Treatment of Water and Wastewater undertook the compilation of a state-of-the-art summary of need, progress, potential, and problems in the use of radiation energy as a treatment technology of water, wastewater and sludge for commercial civil engineering practice. This report discusses radiation sources with energy at or above those of ultraviolet light for use in appropriate aspects of water, wastewater and sludge treatment. The report also identifies some of the areas where radiation energy can be used in synergy with other water treatment techniques, identifies some of the technical issues which would enhance its acceptance, and examines some of the economic constraints to wide-spread application.

Section 2. FUNDAMENTALS OF RADIATION TREATMENT

Radiation treatment may be defined as the application of ionizing radiation energy to a material in order to obtain some useful change in the material.

Ionization

Radiation is energy that travels through space in the form of high-energy particles or electromagnetic waves, such as radio waves, infra-red radiation, and light. Ionizing radiation is that part of the electromagnetic radiation spectrum that possesses sufficient energy to remove bound electrons from the atoms in matter through which it passes. The energy of electromagnetic radiation may be given in terms of temperature T ($E=kT$), where k is the Boltzman constant), in terms of wave length ($E=hc/$), where c is the speed of light and h is the Planck constant), but most generally in terms of frequency ($E=h$). Ionizing energies are of the order of a few electron volts (eV) for some light elements, such as carbon, hydrogen, and oxygen. The ionizing radiation spectrum consists of ultraviolet radiation, x-radiation, and gamma radiation in order of increasing frequency and energy.

Radiation Dosimetry

Radiation dosimetry is the measurement of the radiation energy absorbed in material exposed to a radiation field. The amount of energy absorbed per unit mass (the absorbed dose) depends on three principal factors:

(1) the nature of the material; its chemical and physical state
(2) the nature of the radiation; its type and energy distribution
(3) the energy absorption mechanism; the probability of an interaction, which affects the linear energy transfer (LET).

The original unit of ionizing radiation energy absorption was the Roentgen, defined in 1928 as the amount of energy absorbed by 1 cm^3 dry air at standard temperature and pressure which produces one electrostatic unit of charge of either sign. This unit was replaced by the rad, defined as the absorption of 100 ergs per gram (cgs) or 0.01 J/kg (SI). The present unit of absorbed dose in SI is the Gray (Gy), defined as 1 J/kg.

Radiation can be measured by several types of dosimeters, each of which has a calibrated measure of effect. These include electrical chambers that measure the number of ions or electrons produced; thermoluminescent (crystal) materials that produce light in thermal relaxation; calorimeters that integrate the heat of thermal relaxation; and chemical indicators that have stoichiometric response between absorbed dose and reaction products.

A second important measurement parameter in radiation treatment is the yield of the effect on the material by the radiation interactions. The parameter is called the G-value, and it expresses the amount of a species involved in the irradiation treatment per unit of mean radiation energy absorbed by the material (e.g., in atoms/100 eV or mol/J). For materials in solution, the yield is sometimes given as a Y-value, the change in concentration per unit of absorbed dose (e.g., in μmeq/ml/rad). For example, for the chloroform chemical dosimeter, which measures produced [H^+] with an acid dye, from the measured change in color in a solution density of 1.5 g/cm^3, the Y-value is 0.1 μmeq/ml/rad, and the G-value is 73.6 molecules/100 eV.

Radiation Effects in Bulk Material

The effects of irradiating bulk material with ionizing radiation are due to a large number of individual radiation-atom/molecule interactions. The principal interaction is ionization, described by

$$A + h\nu = A^+ + e^- + h\nu'$$

in which atom A absorbs sufficient energy from the interacting radiation to liberate one of its bound electrons and become a charged ion. The energy-degraded radiation can continue to ionize other atoms in its path. For a mean ionization potential of about 35 eV, a radiation particle (photon) of 3.5 MeV could ionize about 100,000 atoms before it comes to rest. If an interaction results in absorption of only partial ionization energy, the process of excitation may occur, given by

$$A + h\nu = A^* + h\nu'$$

where the * denotes atom A in an excited energy state with one electron in a higher (less bound) orbit.

Following an ionization event, the ionized atom combines with a free electron to an excited state, and the excited atom returns to its original ground state through one of several processes of relaxation, for example, by fluorescence

$$A^* = A + h\nu''$$

where $h\nu''$ is the energy released by the return of the electron from an upper orbit to its fully bound orbit. Another process is charge transfer

$$A^* + B = A + B^*$$

where B becomes an excited atom or molecule. A third important relaxation process is the rupture of covalent bond molecules

$$(R:S)^* = R. + .S$$

where R. and .S are free radicals that have large chemical reactivity energies.

Free radicals play an important role in the processes of radiation treatment. A review of free radicals in radiation chemistry is given in Spinks and Wood (1976). Free radicals undergo a short but active life from their formation, given by the previous equation, by interacting with neighbouring atoms and molecules in a variety of ways:

rearrangement	AB. = BA.
dissociation	AB. = A. + B
addition	R. + >C=C< = >C-C<
substitution	A. + BC = AB + C.

4

until the free radical energy is dissipated in a termination process, such as

    combination   R. + .S   = R:S
    disproportionation   2 RH.   = RH$_2$ + R
    electron transfer   M(z$^+$) + R. = M((z+1)$^+$) + R$^-$

Radiation Chemistry of Water

An important example of the life cycle of free radicals is in the irradiation of water. A review of the radiolysis of water is given by Thomas (1969), in which primary processes, free-radical yields, and the effects of LET and pH are examined. The formation of free radicals is given by reactions as

$$2 H_2O + h\nu = H_2 + 2 \cdot OH$$
$$\text{or } H_2 + H_2O_2$$
$$\text{or } 2 H. + H_2O_2$$

These free radicals can undergo a series of chain reactions, such as

$$.OH + H_2 = H_2O + H.$$
$$H. + H_2O_2 = H_2O + .OH$$

and finally terminate by a combination of reactions as

$$H. + O_2 = HO_2$$
$$.OH + H_2O_2 = H_2O + HO_2$$
$$HO_2 + HO_2 = H_2O_2 + O_2$$

An important factor for commercial applications of radiation energy treatment is the extent of free radical formation and the propagation through free-radical chemical chain reactions. Most G-values lie between 1 and 10. Higher values usually indicate a chain reaction. In water, the G-value for free-radical formation varies with such factors as pH, dissolved oxygen, and the radiation type and energy. Table 1 (from O'Donnell and Sangster, 1970) shows a range of measured yields of primary species from irradiation of pure water as a function of pH at low LET.

Table 1[*]
G-values in Radiolysis of Water at Low LET

Chemical Specie	G-Value 0-2	(molecules/100 eV) 4-11	at pH 13-14
e-(aq)	3.65	2.7	3.1
H.		0.55	0.54
.OH	2.95	2.8	2.9
H_2	0.45	0.45	0.45
H_2O_2	0.8	0.7	0.7
H_3O^+		3.6	
OH$^-$		≈1	

[*] from O'Donnell and Sangster (1970).

Aqueous Radiation Chemistry

Radiation treatment of water supplies and wastewaters implies radiation chemistry of dilute solutions (with contaminant concentrations much less than 1 molar). For such solutions, the primary interactions of the radiation field are mostly with water molecules. These interactions, as in pure water, produce such species as $H_2O^+ + e^-$, H_2O^*, $H\cdot + \cdot OH$, $H_3O^+ + \cdot OH$, HO_2, and H_2. Any of these radicals can interact with solutes or particles in the water as secondary reactions. The two key parameters are reaction times and types. Reaction time can be expressed as a rate constant, k, in units of 1/mole-sec, or as a time constant, τ, where $\tau = kC$, in units of reciprocal seconds.

Reaction types vary primarily with $\cdot OH$ radicals or $H\cdot$ or solvated electrons. The reaction types and approximate rate constants for $\cdot OH$ are represented by

recombination	$\cdot OH + \cdot OH = H_2O_2$	$k \approx 4 \times 10^{10}$
addition	$\cdot OH + \theta = \theta < \dot{O}H$	4×10^9
substitution	$\cdot OH + C_2H_5OH = >\dot{C}-\dot{C}<$	1×10^9
charge transfer	$\cdot OH + Fe^{2+} = Fe^{3+} + OH^-$	2×10^9

Similar reactions for solvated electrons and $H\cdot$ are

$e^-(aq) + H_3O^+ = H\cdot + H_2O$	$k \approx 2 \times 10^{10}$
$e^-(aq) + O_2 = O_2^-$	2×10^{10}
$e^-(aq) + \theta-Cl = \theta\cdot + Cl^-$	5×10^8
$H\cdot + O_2 = HO_2\cdot = O2^- + H^+$	2×10^{10}
$H\cdot + C_2H_5OH = >\dot{C}-\dot{C}<OH + H_2$	2×10^7

Radiation in Water and Wastewater Treatment

The use of radiation in treatment of water, wastewater and sludge is based on achieving a sufficient absorbed dose uniformly throughout a large flow rate to provide an economic yield of the desired treatment effect. The design of an irradiation facility must take into account the type of radiation to be used, its energy distribution, its penetrability into the material being treated, and the geometry of the radiation interaction volume, which defines the thickness of water normal to the radiation flux. The absorbed dose per unit mass is the dose rate in that mass integrated over its total time of irradiation.

The types of radiation useful for water treatment are ultraviolet radiation, x-radiation, which may be available as bremsstrahlung from high-energy particle accelerators, gamma-rays from radioisotope sources, and high-energy electrons from electron accelerators, which provide a mono-directional beam of high-energy electron particles.

The more promising uses for radiation energy in water treatment include

(1) disinfection (inactivation of pathogenic organisms)
(2) bond rupture (destruction of organic pollutants)
(3) oxidation of organic pollutants
(4) enhancement of sedimentation and flocculation processes
(5) enhancement of filtration processes (attainment of larger particle size and improved adhesion).

One of the issues in radiation disinfection of water is in the radiosensitivity of the various types of parasites, bacteria and viruses. A review of virus radiosensitivities is given by Lemke and Sinskey (1975). Many experimental efforts are performed with E.coli as an indicator of the efficacy of disinfection by ionizing radiation. Bryan (1984) notes that neither the ratio of numbers of coliforms to pathogens nor their comparative resistance to disinfection processes is well known. Experimentally measured doses for a 99 % kill of coliforms are of the order of 2×10^4 rad (0.2 Gy). For spore-forming bacteria, the doses required are of the order of 5×10^5 rad (5 Gy), whereas complete sterilization requires doses of the order of $3-4 \times 10^6$ rad (30-40 Gy). Thus the size of the radiation source and the water flow rate, which affect the specific cost of treatment, need to be matched to the degree of disinfection desired.

Another problem is the yield of the irradiation process, both by itself and in combination with other agents, such as chlorine or ozone. G-values for oxidation have been measured in aerated waters. For water without dissolved oxygen, the measured G-value is about 3. In aerated water, with the reactions $O^2 + H. = HO_2$ and $O_2 + e^-(aq) = O_2^-$, the G-value is about 15.

Section 3. GAMMA RADIATION

Gamma rays are emitted during the decay of some radioactive atoms as a process for releasing excess energy from an excited nucleus. Gamma rays have good penetrating power and high probability of interacting with atoms of the material through which they pass. They are used in cancer treatment to kill the cells of tumors, in the sterilization of medical supplies and pharmaceuticals, and in the disinfestation and preservation of food.

The radioisotope used predominantly for industrial irradiation applications is cobalt-60, deliberately produced from elemental cobalt-59 by irradiation with neutrons in a nuclear reactor for periods of one or two years. Cobalt-60 atoms disintegrate with emission of two gamma rays with energies of 1.17 and 1.33 MeV. The cobalt-60 source decays with a half-life of 5.26 years, and thus the source needs to be replenished periodically. A total of approximately 250 MCi of Cobalt-60 are used in more than 140 full-scale, industrial irradiators at work around the world (Malkoske and Gibson, 1990).

Another radioisotope which has been considered for radiation treatment is cesium-137, which is a by-product of the fission of uranium-235 fuel in nuclear reactors. Six atoms of cesium-137 are produced in the fuel for each 100 fissions. Cesium-137 atoms emit one gamma ray of 0.66 MeV per disintegration, but the source has a half-life of 30 years, which requires less frequent replenishment than cobalt-60 sources. The U.S. Department of Energy has placed a moratorium on the use of cesium-137 in industrial irradiators and the future availability of these sources is therefore uncertain.

Advantages of Gamma Irradiation

Gamma rays have excellent penetration power and are capable of providing uniform dosage in materials. The continuous and predictable emission of gamma rays from the source assures availability of power and reproducible dose profiles. Radiation energy is environmentally clean - gamma irradiation systems do not require the use of chemical additives and neither toxic chemicals nor residual radioactivity are produced in the material (Chuagin, et al, 1990). In the operation of a gamma facility, for a given flow rate, source strength, and system design, the only critical operating parameter is exposure time. Energy consumption is generally small, since the irradiation energy is supplied by the gamma source.

Disadvantages of Gamma Irradiation

Some disadvantages of gamma irradiation are:

gamma-ray sources from radioactive materials cannot be switched off when not in use, unlike machine produced radiation;

production and supply of cobalt-60 sources is currently characterized by a small, but increasing, number of suppliers;

irradiation facilities have relatively high capital costs compared to more conventional water treatment technologies, such as chlorination, but are competitive with other methods for dewatered sludge disinfection (see e.g., Sivinski and Ahlstrom, 1984; Markovic, 1986).

Determination of Source Requirement

Factors that affect the strength of a gamma source for water treatment include the source-to-water geometry, the amount of solids in the water, the bulk density, and irradiator design (McKeon, et al., 1983). These factors are not entirely independent of each other. For example, the solids content affects the rheology of the water, wastewater or sludge, and these characteristics influence the type of conveyor required to transport the material into the irradiator. The type of conveyor, in turn, impacts irradiator design and cost.

Gamma sources can be used for municipal wastewater sludge disinfection since gamma rays provide the penetrating capability to process high-density materials. Gamma rays are less suited to treat fast-moving thin streams of water or wastewater. The solids content is a major variable in design of gamma irradiators for sludge. The mass of sludge to be irradiated also depends on the degree of dewatering before irradiation.

Generic Facility - Geiselbullach

The use of gamma radiation for sludge treatment is described by examining the operation of the Geiselbullach facility in Germany, operated since 1973 as the world's longest operating full-scale sludge irradiator.

Studies reported by Lessel, et al (1979) indicated that one economical option for disposal of large quantities of liquid wastewater sludges (4% solids) would be to spread it on agricultural land as a soil conditioner and fertilizer. Two requirements for land application of liquid sludge on grassland and areas used for field fodder cultivation set by German regulations are: the concentration of certain heavy metals in the sludge which must be below prescribed limits, and the sludge must be treated with a state-approved process to render it non-infectious.

Based on research funded in Germany in the early 1970s, Lessel et al, (1979) identified gamma irradiation as an alternative method for disinfection. The method was selected for test trials and a gamma sludge irradiation plant was constructed at Geiselbullach near Munich, which began operation in July 1973. Experimental protocol mandated that all results from the gamma tests be compared against results from alternative conventional methods of disinfection accepted for sludge use on land.

At the conclusion of the demonstration program, a comparison of costs and technical results convinced the operators of the Geiselbullach treatment plant to continue using the gamma facility as an integral part of their sewage treatment and disposal system. The Geiselbullach facility is still a fully operating facility, without special public financial support. The costs of irradiation and land application are borne by the municipality.

At the Geiselbullach wastewater treatment plant, raw sludge is anaerobically digested to stabilize it prior to irradiation. It has a solids content of about four percent and is contaminated with bacteria, parasites, and viruses. The composition (e.g., pH, organic components, trace elements) of the digested sludge is relatively unaffected by irradiation. Following irradiation disinfection, the liquid sludge is thickened by sedimentation and then (8% solids) applied to farmland by tanker-type trucks as a fertilizer and soil conditioner (Lessel and Suess, 1984).

A schematic of the Geiselbullach gamma facility is shown in Figure 1. The facility consists of a building above ground that houses a metering silo, the monitoring and control equipment, an overhead crane, and a small laboratory. A shaft below ground level contains the gamma sources. Another shaft below ground level, adjacent to the irradiation shaft, contains the recirculation and evacuation pumps (Lessel, et al, 1979).

The initial gamma source was about 120 kCi of cobalt-60, providing a dose of 300 krad for treatment of 30 m^3 per day. Augmentation of the source has occurred periodically, including a loading of cesium-137 capsules in 1983. In 1990, the cobalt-60 source loading provided the plant with a capacity of about 130 m^3 of sludge per day. The sludge is irradiated in batches. A sludge volume of 5.6 m^3 is fed from the metering silo down into the enclosed space in the irradiation shaft. The sludge is then circulated around the gamma source to homogenize the dose. The preset irradiation time required for each batch is derived from the desired average dose and the source activity.

The gamma sources are inserted between the double walls of an annulus centered in the irradiation shaft. Primary shielding is achieved with poured concrete around the sides and at the bottom of the shafts, and a 1.8 m thick concrete cap at the top of the irradiation shaft, which is the only restricted area. Instrumentation is located in the pump shaft and in the building above ground. Service and inspection of components are thus made at any time without special safety precautions. The automated plant operates 24 hours per day, and has averaged approximately 349 days per year of continuous operation.

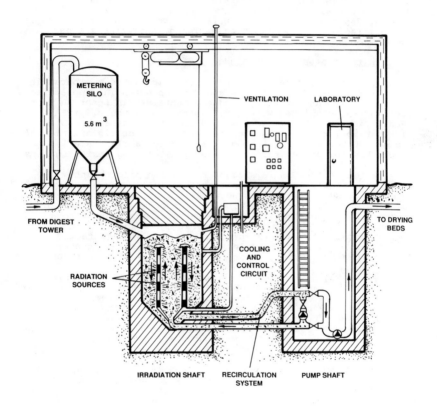

Figure 1. Schematic of the Geiselbullach Gamma Sludge Irradiator
(from Lessel, et al, 1979).

Other Facilities

In addition to the Geiselbullach liquid sludge irradiator in Germany, a number of gamma-irradiation facilities have been installed around the world to treat wastes, especially municipal wastewater sludges. In the United States, an experimental sludge irradiator has operated successfully at the Sandia Laboratories in Albuquerque, New Mexico. Several other countries, including Canada, Japan and India have operated experimental facilities for waste irradiation investigations. These facilities are described in Section 8.

United States - Dried Sludge Irradiator

The U.S. Department of Energy (DOE) initiated a program in the early 1970s to investigate and develop potential applications for cesium-137 and other by-products contained in spent nuclear fuel wastes associated with a determination by the U.S. EPA (Regulation 40 CFR 257) that a dose of 1 Mrad was an accepted alternative for obtaining the required degree of disinfection of sludge. Other criteria that determine recyclability are acceptably low concentrations of heavy metals and toxic organics.

Design of the Sandia sludge irradiation pilot plant (Sivinski, et al, 1983) was initiated in 1977. The plant was intended to serve several purposes: (1) provide a suitably-sized research facility for evaluating key components to be used in larger irradiation facilities, (2) serve as a precursor to a commercial facility by providing operating, economic, and environmental data, and (3) produce a source of disinfected sludge for research. The pilot plant was dedicated in October 1978, as the Sandia Irradiator for Dried Sewage Solids (SIDSS).

In the initial phases of the program, the design concept for the irradiator was based on obtaining synergistic effects of heat and radiation to disinfect liquid sludge (up to 6 percent solids content). However, further cost analysis showed that gamma radiation was better suited for irradiation of dried, bulk material and project emphasis was shifted in that direction. The pilot plant was designed for continuous irradiation of sludge dewatered to a solids content of 40 percent or larger.

An isometric cutaway view of SIDSS (from Ahlstrom, 1985) is given in Figure 2. The irradiator is comprised of two concrete-shielded vaults: a source handling pool and an irradiation chamber. These two areas adjoin one another along a common wall. The gamma source consisted of 15 capsules of cesium-137 produced at the DOE Waste Encapsulation and Storage Facility (WESF) in Hanford, Washington. The total source strength was slightly under 1 MCi, equivalent to about 125 kCi of cobalt-60. For irradiation, the source capsules were loaded into the source plaque at the bottom of a 20-ft stainless steel pool, the pool covers were replaced, and the water was drained. The large lead shutter was retracted to allow movement of the plaque from the pool to the radiation area.

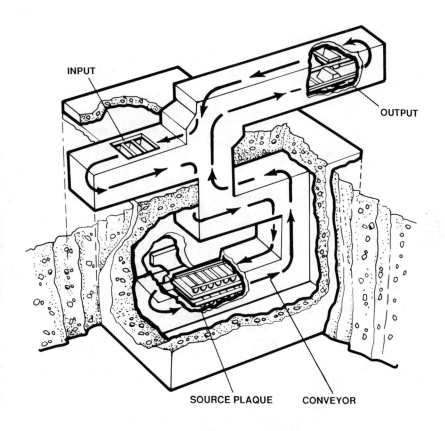

Cutaway of Sandia Irradiator
for
Dried Sewage Solids.

Figure 2. Isometric view of SIDSS, Albuquerque, NM (from Ahlstrom, 1985) .

Sludge, either bagged or in bulk, was transported into, through, and out of the radiation compartment in a bucket conveyor. The conveyor buckets made one pass above the horizontal source plaque, and one pass below the plaque. Details of the operation of the facility were described by Ahlstrom (1985).

The SIDSS pilot plant was operated intermittently for approximately 6 years. A summary of the plant operations was reported by Sivinski and Ahlstrom (1984). Measurements and calculations of absorbed dose performed at the facility were reported by Morris (1980). In 1985, the irradiator completed its research mission, and the cesium-137 source was removed from the facility.

Canada - Partially-Dewatered Sludge Irradiator

In 1987 the Canadian Federal Government, through its Crown Corporation, Nordion International Inc., began investigating the sludge irradiation option. Results of these technical, economic, and process feasibility studies have been positive and have encouraged plans for the installation of a Canadian sludge irradiator. The initial phase of the project plan (Swinwood and Wilson, 1990) has emphasized research, testing, and product development trials, along with system design and capital and operating cost analysis. The second phase of the project plan calls for the construction and operation of a full-scale demonstration Sludge Disinfection System, including a sludge irradiator, at a Canadian wastewater treatment plant. The final phase of the project will see the facility turned over to city staff for commercial-scale operation.

The demonstration Sludge Disinfection System (SDS) facility, shown in Figure 3, will be housed in a warehouse-like building (approximately 2,000 m^2) integrated into an existing municipal wastewater treatment operation (Swinwood and Wilson, 1990). The SDS consists of four main components : 1) sludge dewatering equipment; 2) gamma irradiator; 3) sludge conveying equipment; and 4) truck loading mechanism. The sludge dewatering equipment will dewater the sludge from 3% solids (a thick soupy consistency) to about 25% solids (wet earth consistency). The dewatered sludge will be metered into large aluminum boxes (totes), which will move on a conveyor mechanism, into the concrete-walled irradiation room for disinfection. Disinfection will be achieved by cobalt-60 gamma radiation. Following disinfection, the totes will be moved out of the irradiation room to an unloading station, where the sludge will be dumped into trucks for transport to a fertilizer production company. The finished product will be an organic, sanitary fertilizer/soil conditioner.

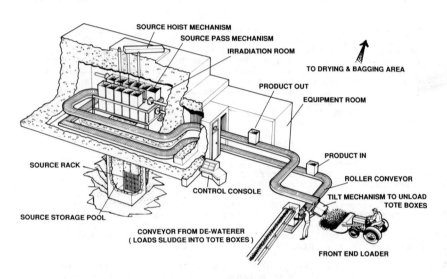

Figure 3. The Canadian Sludge Disinfection System
(from Swinwood and Wilson, 1990).

The gamma sludge irradiator will consist of three major components : (1) the irradiation chamber including source storage pool; (2) the product handling mechanism including control and safety sub-systems; and (3) the cobalt-60 gamma radiation source including the source movement mechanism. The irradiation chamber serves two primary functions, providing safe shielding of the radiation source and serving as the irradiation chamber. The irradiation chamber is a concrete room and a water-filled pool. Its internal dimensions are 7 x 7 m and 3 to 6 m in height. The concrete thickness varies with maximum source strength, but is typically 1.8 m thick. The water-filled pool, 6-9 m deep, provides safe storage of the cobalt-60 source when the system is not in use.

During radiation processing, the source is raised by pneumatic hoists from the storage pool into the irradiation chamber. The product is conveyed by an electro-mechanical, pneumatic and/or hydraulic handling mechanism in a predetermined pattern around the source. The cobalt-60 source consists of a vertical, stainless steel frame or plaque, 2 to 3 m wide, 2 to 5 m high, by 2.5 cm thick, which holds a number of stainless steel sealed tubes approximately 45 cm long by 1 cm diameter, each containing a specified amount of cobalt-60.

Section 4   ELECTRON-BEAM RADIATION

High-energy electrons are available either as beta radiation from radioactive atoms or from production in high-voltage accelerators. Radioisotopes, such as cobalt-60 or cesium-137, emit beta radiation with electron energies in the range 0.1 to 1.5 MeV, generally considered small for large-scale radiation treatment of water. Electron accelerators produce "energized electrons" with energy to a practical limit of about 10 MeV. The electrons are collimated into a beam and accelerated through the high-voltage electric field to provide the desired energy flux.

The relative energy efficiency of the electron-beam for disinfection is difficult to compare with the energy efficiency of other means, such as chemical (e.g., halogen) disinfection or thermal disinfection (pasteurization). The problem stems from the difficulty of estimating energy consumption for producing chemical disinfectants and temperature requirements. Thermal sterilization requires a temperature increase of 100 °C which, for water and aqueous suspensions, can be accomplished only under pressure twice that of normal atmospheric pressure. Some of that energy can be recovered by use of heat exchangers to preheat incoming fluid at increased capital and operating expense. Absorption of a disinfecting dose of 400 krad (Metcalf, 1979) results in a product temperature increase of only 1 °C.

A description of an electron accelerator is given by Schuler (1979). Two types of high-voltage accelerators are currently in use (1) electrostatic (such as the Van de Graaf accelerator) and (2) oscillating (the linear accelerator). Electrostatic accelerators have charge carriers to provide the required high voltage. Linear accelerators have power oscillators that provide very high frequency voltage. The electron accelerator generally consists of: (1) an electron source (usually a glowing filament); (2) an acceleration zone (usually a static or high-frequency electric field) and (3) a horn or output area that shapes the beam. The system is enclosed under constant vacuum. The electron beam is oscillated (scanned through the product) to provide a uniform irradiation dose.

Electron beams have a different interaction geometry compared to gamma radiation. Gamma radiation from material sources is emitted isotopically and efficient use of the gamma source requires conforming to this property. In contrast, electrons can be focused directionally as a beam and can scan horizontally and vertically. The limited penetration capability of an electron beam limits irradiated water thickness. Although the relationship of dose to dose rate is not fully understood, the ability to apply a high dose rate makes the electron beam especially advantageous for inducing chemical reactions that are rate-dependent, such as in decontamination of water or destruction of hazardous organic substances.

Electron-Beam Treatment

Sludge Disinfection

Research in the United States on the use of high-energy electrons for sludge disinfection was initiated in 1974, as reported by Trump (1975). The original project objective was to study the potential use of an electron beam for disinfection of domestic wastewater. About the same time, results reported by Malina, et al (1974) showed that viruses in domestic wastewater were concentrated in sludges derived from the treatment process. The adsorbed viruses survived subsequent sludge processing including anaerobic digestion. When the sludge was applied to soil, viruses initially adsorbed to sludge and soil particles remained viable for long periods of time and were capable of being released while still viable under conditions simulating rainfall. Moore et al (1976) reported on virus survival in Colorado soils, especially under cooler temperatures and ideal soil conditions.

As a result of concerns about reducing sludge-handling personnel exposure to pathogens and the possibility of pathogen contamination from sludge applied to land, further study was addressed to the use of energized electrons for sludge disinfection. Sinskey, et al (1976) and Metcalf (1979) reported on the energy efficiency of electron beams for virus inactivation. Metcalf (1979) noted that a disinfection dose of 400 krad was adequate to ensure complete disinfection of viruses in digested sludge.

Guymont (1978) reviewed disinfection processes for compost considered suitable for public distribution of the product. These included gamma and energized electron radiation, pasteurization and long-term storage. He found a comparison of the options to be difficult because little work has been done on disinfection of compost and the processes suggested were for the most part in the developmental stage. Kirkham and Manning (1979) examined the effect of electron beam irradiation of sludge on its capability to provide plant nutrients to vegetables. They concluded that the irradiation had no effect on plant nutrient availability nor uptake from either raw or digested sludges.

Water Treatment

Trump, et al (1976) examined the potential of electron beam capability to destroy trace amounts of polychlorinated biphenyls in water solution with relatively small doses of radiation. He showed that 95% of the 4-chlorobiphenyl present at a concentration level of 0.8 mg/l in water was destroyed by a dose of 10 krad and virtually eliminated with a dose of 30 krad. Merrill (1987) reported on electron-beam decontamination of water containing aromatic hydrocarbons in the presence of polymers. For electron-beam irradiation in the presence of polyethylene oxide, a number of reaction products disappeared while the concentration of others was reduced drastically. He suggested that free radical species formed by the attack of hydroxyl radicals on the organic contaminant were coupled to radical sites formed on the dissolved polymer, resulting in removal of the organic solute which in turn could be removed from water by conventional sand filtration. Waite, et al (1990) reported on the removal of trace concentrations of ethyl chlorobenzene from water and wastewater. The possibility was raised that electron-beam irradiation could be used effectively for decontamination of groundwater.

Current Status of Electron-Beam Technology

A number of electron-beam studies on treatment of wastewater and municipal sewage sludges have been conducted at several facilities around the world. Two of these in the United States were the Deer Island plant in Boston and the Miami-Dade facility in Florida. Several other facilities are described in the IAEA (1975) Proceedings and a recent review of current efforts is given in IAEA (1990).

The Deer Island Treatment Facility

The basis for the design and experimental operation of the large-scale electron beam unit installed at the Deer Island Wastewater Treatment Plant originated from Trump (1975). Bryan (1990), in reviewing developments through 1980, noted the results that led to the decision to evaluate the electron accelerator as an alternative to heat treatment for sludge disinfection. Trump, et al (1976) presented three options for sludge management incorporating the use of electron beam disinfection. One of these bypassed the digestion process, i.e. direct injection into soil following disinfection. This method conserves the nitrogen content of the sludge, a factor that would offset disinfection costs if nitrogen conservation were a significant factor. As an incremental step in a conventional system for sludge management involving digestion, dewatering and composting, they cited a projected cost of about $2 per dry ton, based on a disinfecting dose of 400 krad determined by Sinskey (1976) as being adequate to inactivate viruses in wastewater or digested sludges at Deer Island. For a 400 krad dose, the Deer Island unit processed 100,000 gallons of sludge per day containing 5% solids and required 4.5 kWh per wet ton of sludge.

The Virginia Key Wastewater Treatment Plant

Electron-beam treatment was selected as an alternative to heat treatment for sludge disinfection at the Miami-Dade Sewer Authority Department's Virginia Key Wastewater Treatment Plant in Miami, FL. Figure 4 shows a schematic view of this facility. The unit was designed to provide a dose of 400 krad to 170,000 gallons of sludge per day containing from 2 to 8 percent solids. The beam was designed to penetrate 5 millimeters of water at a rated acceleration voltage of 1.5 million volts. The power supply output was 75 kW. The transformer and the accelerator tube are housed in a 2000 cu. ft. Insulated Core Transformer (ICT) tank insulated with sulfur hexafloride gas. The accelerator tube exits in the beam room. The electron beam scanner (horn) is surrounded by 4 ft concrete walls and contains a 4 ft weir over which the irradiated fluid flows. The electron beam penetrates the thin sheet of wastewater or sludge as it flows over the weir and before it falls into outgoing troughs. Electrons that emerge from the fluid sheet are reflected back into the stream by a parabolic reflector.

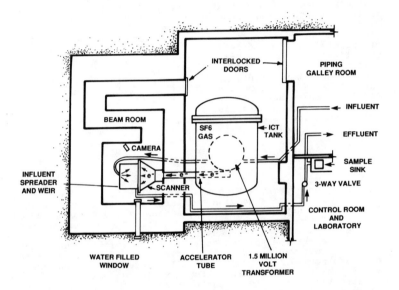

Figure 4. Schematic of the Miami-Dade Electron Beam Accelerator
(from Morse, 1989).

Fergen and Cowgill (1985) noted that the Miami-Dade unit operated in 1984 for one year to meet design performance specifications and to examine electron-beam disinfection in expansion of the Authority's sludge management capability. The results of inactivation performance showed that the electron beam consistently reduced the numbers of fecal coliform and salmonella to below detection limits from influent counts in excess of 100,000 per millilitre for fecal coliforms and about 2,000 per millilitre for salmonellae. During the studies, it was also noted that the normal treatment pathway, especially two-stage anaerobic digestion, was equally effective in inactivating viruses.

Waite (1990) reported on the current research on electron-beam decontamination of water and wastes at this plant, refurbished in 1988. The studies include domestic wastewater as well as secondary effluent to determine the efficiency of electron-beam treatment of such streams.

Research Efforts in Japan

Kawakami, et al (1981) and Hashimoto, et al (1988) have been studying the composting of radiation disinfected sludge for agricultural reuse. Their research suggests that irradiated sludge composted at 50 °C is ready for use after three days, as compared to unirradiated sludge which takes about 2 weeks. Takehisa et al have also examined city water irradiation and Arai and Machi (1988) examined the irradiation of sludge supernates.

Research Efforts in Canada

Research at the Atomic Energy of Canada, Ltd.'s Whiteshell Nuclear Research Establishment in Pinawa, Manitoba has included studies of electron-beam irradiation of liquid sludges and other wastes. The AECL plant uses a 10 MeV, 1 kW prototype industrial linear accelerator designed for research and pilot-scale irradiations. Hare (1990) described the characteristics of the machine. Research has been focused on pathogenic microorganism inactivation and detoxification of organic chemicals such as PCB's and other halogenated organic compounds.

Research Efforts in Austria

Getoff and Solar (1988) at the Institute of Theoretical Chemistry & Radiation Chemistry of the University of Vienna have investigated the decomposition of organic pollutants, such as phenols, in drinking water, both with and without the presence of oxygen. Gehringer, et al (1989) at the Austrian Research Centre has also been carrying out a study to remove perchloroethylene and trichloroethylene in drinking waters by means of electron-beam irradiation. They found that for pollutant concentrations below about 1 ppm, the pollutant decomposition follows a first-order rate law. The dose necessary for removing 90% of perchloroethylene was 255 Gy and for trichloroethylene it was 120 Gy.

Research Efforts in France

In 1979, Levaillant et al carried out an extensive study of chemical and biological effects of water treatment with high-energy electrons. Elimination of bad flavours in raw water supplies was an interesting result of this work.

Potential Electron-Beam Applications

Development of electron-beam applications requires an acceptable basis for direct comparison of the radiation process with the competing processes. For disinfection, a common standard for determining performance is lacking. Bryan (1984) noted that much of the research on disinfection processes has been based on the impact of the process on indicator organisms rather than on the inactivation of potentially harmful organisms, the actual target of the disinfection process.

An integrated process combining electron beam disinfection of sludge and injection into soil that emerged from hypothetical combination of the electron-beam and sludge injection concepts proposed by Bryan (1980) was estimated to cost $30 per dry ton, approximately equal to the cost cited by Epstein and Wilson (1974) for use of the Beltsville method of composting vacuum filtered, digested sludge. Additional economic advantage would result from the use of the integrated process because its use would make unnecessary the costs of thickening, digestion, elutriation, chemical and physical conditioning, and physical dewatering processes such as centrifugation, pressure and vacuum filtration. Pipeline transport of sludges from the point of their origin to their final location as a constituent of topsoil at a site dedicated to sludge management would also reduce the nuisance, risks and costs associated with alternative methods of transporting sludges to their final location (Bryan, 1978).

Section 5   X-RAY APPLICATIONS

X-rays, the energy released by orbital electrons on de-excitation, can be produced in several ways. Conventional x-rays, such as are used in medical diagnosis and therapy, are produced by impinging electrons on a target in an x-ray machine. Synchrotron x-rays are produced by centripetal acceleration of electrons in a high-energy accelerator (Batterman and Ashcroft, 1979). Conventional x-rays, (Gaudin and Fuerstenau, 1960) and synchrotron x-rays, (Bierck et al., 1988) have been used for non-destructive testing to extend fundamental understanding of treatment processes. Application of x-rays in areas such as medical diagnosis and cancer therapy is well established, but x-ray energy has not been applied, insofar as is known, to treatment of water and wastewater.

An x-ray irradiation facility for water or wastewater treatment would resemble an electron irradiation facility with one major change, the introduction of an electron target containing atoms with high atomic number. Interaction of electrons with the target produces x-rays. Conversion of energetic electrons into x-rays results in increased penetrating power of the radiation, but irradiation costs to achieve a given dosage would increase substantially because the conversion efficiency is low.

<u>Current State of Technology</u>

Although Ballantine, et al. (1969) did not include x-rays in an evaluation of radiation energy sources for wastewater treatment, in a later paper, Ballantine (1975) noted that x-rays could achieve better penetration than the electrons from which they are produced, but observed that only about 6.5 percent of energy from a 3 MeV electron accelerator could be converted to x-rays. Trump (1981) indicated a conversion efficiency of about 7 percent, and concluded that the low efficiency of conversion from electron beam power makes x-rays uneconomical for environmental applications involving large volumes of material.

Industrial irradiation of food and medical supplies is more highly developed than irradiation in water and wastewater treatment. In those applications, energetic electrons and gamma rays from radioisotopes are used. High-intensity x-ray generators were not used in these applications as of 1987 (Cleland and Pageau, 1987a). However, various analyses [e.g., by Farrell, et al. (1983) and Cleland and Pageau (1987b)] indicated that x-rays produced by electron bombardment of a high density target could be competitive with gamma radiation. Details of such a facility, being installed in Japan, have been described by Thompson and Cleland (1989). Cleland (1989) expects an x-ray generation efficiency of about 7 percent. Yotsumoto, et al., (1988) describe a 5-MeV bremsstrahlung x-ray source, planned for sterilizing medical devices in Japan, with a calculated conversion efficiency of 10.8 percent.

McKeown (1983, 1985) estimated that x-ray generation efficiency could be increased to about 20 percent with a 12-MeV beam. However, the U. S. Food and Drug Administration limits maximum energy with which food can be treated with x-rays to 5 MeV in order to avoid exceeding the energy threshold for inducing radioactivity in the irradiated material (Cleland and Pageau, 1987b). Although energy limits beyond 5 MeV have not been set for x-ray irradiation of sludge, Leboutet and Aucouturier (1985) reported only a small increase in induced radioactivity when energy was increased from 5 MeV to 10 MeV.

Potential Applications

To assess the possible role for x-ray radiation in water and wastewater treatment, it is prudent to select the application in which the technology would be the most competitive with alternative sources of radiation energy. This application would be irradiation of dewatered sludge. Of potential environmental engineering applications, dewatered sludge irradiation involves comparatively small volumes, and thus the high cost of x-rays would be least disadvantageous in this application. Irradiation of dewatered sludge also might take better advantage of the greater penetrating power of x-rays relative to electrons.

Measurements of the x-ray absorption coefficient of the solid phase of wastewater sludges are not known, but Bierck and Dick (1988) found the difference in x-ray absorption between water and activated sludge solids to be too small to allow accurate measurement of suspended solids concentrations by observing x-ray attenuation. While x-ray penetration in dewatered sludge thus may not be significantly different than penetration in water, still the difficulty of presenting dewatered sludge in adequately thin layers to an electron beam may represent a potential advantage to x-rays.

Hashimoto, et al. (1988) reported improved microbial kinetics when irradiated sludge was composted. While Hashimoto and his coworkers used electron beam irradiation, the advantage would accrue also to x-ray irradiation of dewatered sludge.

A disadvantage of applying x-rays to previously dewatered sludge is that no benefit would be derived from the beneficial effect of ionizing radiation on the physical properties of sludge solids. Investigators such as Etzel, et al. (1969), Compton, et al. (1970), Holl and Schneider (1975), Seferiadis (1977), and Jordan and Erickson (1978) have reported that ionizing radiation improves release of water from sludges. Groneman and Schubert (1978) attributed the mechanism responsible to oxidation by .OH radicals. An economic analysis conducted by Hasit and Dick (1981) showed that reduction in sludge conditioning and dewatering costs due to ionizing radiation could be significant, but that it was not adequate to justify irradiation for that purpose alone.

The potential for x-ray irradiation of sludges has the same disadvantages of other irradiation processes, (1) the extent of viral inactivation may be inadequate, (2) reinfection may occur, and (3) vector attraction may prove to be a problem. Additional disadvantages of x-rays in comparison to other means for sludge irradiation are that (1) the improvement in sludge thickening and dewatering properties resulting from irradiation would not be realized (because it is assumed x-rays most likely would find application to sludges already thickened and dewatered), and (2) treating sludge with ionizing radiation can be expensive.

One advantage of x-rays compared to electron irradiation is greater penetration. This characteristic could be used most advantageously with dewatered sludge. X-rays may also compete with gamma rays in this application. Although gamma rays traditionally have been used when these two technologies compete in radiation applications in other areas, a large-scale x-ray installation has been recently considered. A significant advantage of x-rays in this competition is that they can be turned off and need not be transported or stored.

As with other forms of treatment using ionizing radiation, the feasibility of x-ray irradiation must be evaluated in comparison with competitive nonradiative treatment methods. Given the inefficiency of energy use in x-ray production and the performance limitations previously described, it is difficult to envisage a significant role of x-rays in water and wastewater treatment. This assessment is in agreement with that of Iverson (1987) who concluded that large scale application of radiation processing in the waste field appears to require a change in economic conditions.

Section 6. ULTRAVIOLET RADIATION

Ultraviolet (UV) radiation is defined as the portion of the electromagnetic radiation spectrum lying between visible light and x-rays. The ultraviolet region of the radiation spectrum, with energy generally expressed in wavelength, occupies the 4 to 400 nm band. This region is commonly divided into three subregions: short wavelength (UV-C), ranging from 180 to 280 nm; medium wavelength (UV-B), ranging from 280 to 320 nm, and long wavelength (UV-A), varying from 320 to 380 nm.

Sources of UV light are divided into two classes: natural and artificial. The sun is the most important natural source of UV light although much of its transmitted energy is in the longer wavelength subregion above 295 nm. The radiant energy received from the sun is responsible for the development and continued existence of life on the earth by photosynthesis. As noted by Schenck, (1979), the oxidizing and germicidal effects of sunlight, to a considerable extent, contribute to the conservation of our environment by natural photochemical processes in the atmosphere and by natural ultraviolet purification of surface waters.

UV can be generated artificially by a wide variety of arcs and incandescent lamps. One common type is the low-pressure mercury vapor lamp, which produces UV as a result of electron flow between the electrodes through ionized mercury vapor. These lamps can supply UV radiation energy in such relatively high dose rates that, in fractions of a second, a higher degree of disinfection can be accomplished than by sun irradiation. The low-pressure mercury lamps convert input electrical energy into UV energy with a wave length of 253.7 nm. It is fortuitous that the maximum UV sensitivity of microorganisms and the UV emission of the low pressure mercury vapor lamp fit well together. For this reason the nearly monochromatic low pressure mercury lamp has prevailed as the dominant radiation source in research and practical applications (Schenck, 1979), although medium-pressure lamps are also in use.

Certain portions of the UV radiation spectrum have been found to have more pronounced bactericidal action than solar radiation. UV radiation with energy of the narrow band lying between 200 and 310 nm has the greatest injurious and lethal effects on microorganisms (Carlson, et al, 1985). The maximum microbiocidal action is at about 260 nm for practically all microorganisms. Figure 5 (from Carlson, et al, 1985) shows the relationship between bactericidal effect and wavelength. The maximum microbiocidal action of UV is essentially congruent with the UV absorption and photochemical sensitivity of deoxyribonucleic acid (DNA). The inactivation of microorganisms is essentially based upon photochemical reactions in the DNA which result in errors and faults being introduced into the coding systems. Nature has developed various means of molecular-biological error correction for the protection of the vital DNA and the selectivity of the reactions may be influenced by changes in the organisms in different phases of their life cycle. Thus, photo repair systems may resuscitate a seemingly dead organism by either longer wavelength photon irradiation or dark incubation.

The germicidal action of UV results from its exposure or direct contact with organisms and can only be effective if it is absorbed. The lethal effect of UV is due to a photochemical reaction initiated by absorption of a photon by the molecular structure rather than through formation of a toxic substance in the medium (Schenck, 1979, Hoather, 1960). The inactivation of microorganisms by UV irradiation is proportional to the radiation intensity (mW/cm^2) multiplied by the time of exposure (sec). The product is called the UV dose. The same exposure can be obtained by using a high intensity for a short time or a low intensity for a proportionally longer period of time (Reddish, 1957, Lawrence and Block, 1968). It should be noted that, with UV radiation, too high a dosage causes no adverse effects except additional cost.

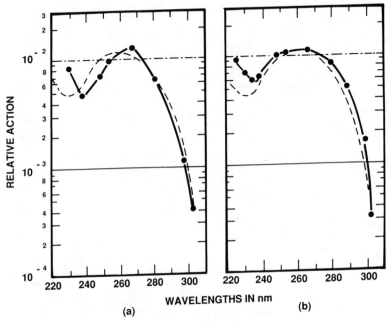

Action spectra for killing of (a) *Escherichia coli* and (b) *Staphylococcus aureus*. The broken lines show the relative absorption of DNA.

Figure 5. Concurrence of DNA absorption of UV energy and germicidal action as a function of wavelength
(from Carlson, et al, 1985)

Application of UV Radiation to Water Disinfection

The first recorded attempt to use UV for public water supply treatment was made in 1910 at Marseilles, France (Walden and Powell, 1911). Later, four municipalities in the United States adopted UV for water treatment during the period dating from 1916 to 1928 (Baker, 1948, Ellis and Wells, 1941). Other installations of UV equipment were made in various parts of the United States and Central America serving industry, hotels, institutions, and ships, but most were abandoned before 1930 (Baker, 1948). The main reasons given for abandoning the UV method of treatment were: relatively high operating costs, operating and maintenance problems, and the advent of chlorination which was found to be more efficient and reliable.

Subsequently, during the decades of the thirties and forties, considerable effort was directed toward the application of UV to the destruction of airborne bacteria and air sanitation (Oda, 1969). With the development of the new tubular germicidal UV lamp, renewed interest in UV water disinfection resulted in considerable experimental data from numerous studies (Luckiesh and Holladay, 1944, Koller, 1965, Buttolph et al., 1953).

However, in spite of the new and more efficient UV equipment, a report in 1948 by the Council on Physical Medicine concluded that UV disinfection was not completely reliable for purification of public water supplies (Council on Physical Medicine, 1948). Subsequent environmental problems and certain limitations associated with chlorination have prompted research into UV for disinfection of potable water and wastewater effluent. Chlorination produces residuals and by-products which may be carcinogenic to humans and which can be toxic to aquatic life in receiving waters. In addition, UV disinfection may be more effective than chlorination in killing viruses (Yip and Konasewich, 1972). The main impetus for increased study of UV disinfection resulted from the dramatic increase in waterborne disease episodes in the United States since the mid 1960s. Specifically, the widespread occurrence of giardiasis, whose etiologic agent, $\underline{G.\ lamblia}$, has been found to be highly resistant to control by conventional chlorination, has led to renewed interest in UV as an alternative method of water disinfection.

In the last decade, the use of ultraviolet light as an alternative to chlorine in the disinfection of wastewater has increased significantly. This change in disinfection source has been pushed by: (1) the negative aspects of chlorination; (2) requirements for dechlorination of wastewater effluent resulting from more restrictive limits on residual chlorine in receiving waters; (3) generation of undesirable chlorinated organics with the use of chlorine; and (4) concerns over the safety of transport and storage of chlorine and sulfur dioxide. Conversely, UV treatment: (1) has no residual chemicals; (2) does not affect chemical reactivity; (3) does not have safety concerns that would affect surrounding communities; and (4) does not significantly affect normal plant safety concerns.

The increased use of UV can be shown by comparison of plant surveys made by Scheible (1985) taken four years apart in 1984 and 1988. In 1984, 53 operating wastewater treatment plants were using UV disinfection. Most were small with 80% having design flows of less than 1.0 mgd. By 1988, there were nearly 300 plants in operation. Of the 177 plants for which data were available, 59% were less than 1.0 mgd. However, larger systems were going into operation with 38% having design flows between 1 and 10 mgd and several at greater than 10 mgd. The largest operating plant, at Madison, Wisconsin, has a design peak over 100 mgd. Two plants at Quebec City will have a design flow of 235 mgd subjected to UV disinfection.

Factors affecting use of UV disinfection of wastewater include: (1) wastewater quality parameters of flow, initial microbial concentrations, suspended solids, and the UV absorbance; (2) hydraulic behaviour of the UV reactor residence time distribution, dispersion characteristics, turbulence, effective volume, flow patterns, and head loss; and (3) UV radiation delivery involving radiation intensity, UV unit design, and UV exposure pathways. Another factor of concern has been the establishment of mechanisms for defining adequate removal of pathogenic organisms.

To be effective as a disinfectant, UV radiation must penetrate the pathogen's cellular machinery and permanently disrupt the living cell's ability to reproduce and to cause infection. The well-designed system must contravene impediments to UV transmission to the cell's interior and assure that sufficient energy does reach the controlling sensitive areas of the cell. Since UV is absorbed by water and particulate matter, disinfection efficiency is enhanced by removal of suspended solids prior to disinfection and by maintaining the shortest feasible pathway between the UV source and the subject pathogens. Surfaces which intercept the UV light path between the UV source and the pathogen must be kept clean so that absorption of the UV radiation does not occur in collecting solids, deposits of dust, debris, and biological growth. The time of exposure to radiation needs to be as uniform as possible. Thus, short circuiting and areas of blinding from the radiation need to be avoided in design.

As more treatment facilities include UV disinfection in their treatment train, the design factors, the UV equipment, and the operational techniques will continue to improve, and there should be greater acceptance of UV as a commendable alternative to disinfection with chlorine. However, since UV provides no residual in the treated liquid, chlorine may continue to be the source of residual disinfection power following the UV process.

Design of UV Reactors

The ability of disinfecting agents to meet established effluent standards is a function of several factors including: (1) the agent's inherent ability to inactivate (biocidal characteristics); (2) mixing (mass transfer) of the disinfectant into the process stream; and (3) the hydraulic dispersion characteristics of the reactor. Of these three factors, the last two are more amenable to control in design. Depending on the degree of disinfection, these factors take on different levels of importance for different disinfectants.

Unlike chlorine and ozone, ultraviolet radiation does not have problems with mixing. Its physical nature renders UV free of the problems of mass transfer. However, like chlorine and ozone, contactor flow dispersion has a significant impact on the ultimate ability of UV reactors to attain significant disinfection levels. This is especially true in light of UV's inability to produce a residual.

Several ultraviolet disinfection studies have shown that, as reactors approach plug flow, greater degrees of disinfection can be attained at lower doses, and therefore, lower costs. In a study by Severin et al. (1983, 1984), a completely mixed reactor could attain only a three-log reduction of E.coli at an average dose of 6820 $mW-sec/cm^2$, although log reductions greater than five were achieved in batch studies in a completely stirred treatment reactor (CSTR) at a much lower dose of 22.5 $mW-sec/cm^2$. Initial resistance of the microorganisms, rather than kinetic inactivation rates, appeared to be the limiting factor in establishing the disinfection limits.

Another study (Petrasek et al., 1980) reported that when the ratio of actual detention time to theoretical detention time (T_a/T_{th}) of a reactor dropped well below 1.0, disinfection was significantly hindered. Using a Kelly-Purdy UV reactor (fluorescent UV lamps suspended longitudinally above a sheet flow), flow depth in the reactor was varied to study the effect of (T_a/T_{th}) on coliform disinfection. For a dosage of 32 mW-sec/cm2 applied to a 2.54 cm depth of flow with a (T_a/T_{th}) = 1.0, total coliform and fecal coliform concentrations were reduced 3.38 and 3.40 logs, respectively. In comparison, for a (T_a/T_{th}) = 0.77, (a dosage of 38 mW-sec/cm^2 applied to a 6.35 cm depth of flow) total coliform and fecal coliform reductions were only 2.29 and 2.25 logs, respectively. The difference, primarily a function of short-circuiting, appears to be expressed by the (T_a/T_{th}) ratio.

Johnson et al. (1982) reported that, when **B. subtilis** spores were used as a tracer for a thin film PWS-UV reactor, most surviving spores emerged from the UV contactor well before the average retention time. This unit exhibited both short-circuiting and tailing in residence time distribution (RTD) curves. (T_a/T_{th}) ratios were 0.75 and 0.79 for flow rates of 2.0 l/sec (32.3 gpm) and 0.6 l/sec (8.84 gpm), respectively. Because of short-circuiting, coliform reductions were limited to approximately 3 logs. In comparison, a second UV unit, which attained greater than 4.5 log reductions in coliforms, had significantly less short-circuiting and tailing. (T_a/T_{th}) ratios were 0.95 and 0.97 at flow rates of 2.9 l/sec (45.5 gpm) and 0.6 l/sec (8.84 gpm), respectively.

These studies demonstrate the need to maintain nearly plug flow conditions in the disinfection reactor to attain effective disinfection. The total coliform maximum contaminant level (MCL) for potable water is so stringent that even small degrees of short-circuiting through the reactor make it impossible to attain the necessary level of disinfection. Compensation for short-circuiting problems can be achieved by increasing the UV energy reaching the microorganisms, although this increase will result in increased operating costs, and may affect its economic competitiveness.

Current Problems in UV Treatment

A major problem in UV disinfection units is the need for improved dosimetry. A self-contained dosimeter to read the actual UV dosage directly is not readily available. The general method is actinometry to gauge the output of the UV unit. Since the geometry of the unit, the flow patterns at various Reynolds numbers, the degree and intensity of reflection, the average distance between lamps, the penetration depth, the cleanliness of the liquid and of the UV lamp, the degree of short-circuiting, and the degree of light scattering all have potential effects on the dosages applied, photo actinometry is used to assess the UV light intensity of exposure.

Conventional radiation dosages are commonly reported as an energy flux (energy/cross-sectional area). Flux measurements can be determined with a radiometer, a photosensitive membrane which reacts to energy flux due to thermal effects. In practical applications, the use of a radiometer is limited by the geometric design and physical scale of the contactor. Direct correlation of ferrioxalate actinometry with radiometer measurements makes it possible to indirectly use radiometer measurement as the dose measurement, expressed in units of $mW-sec/cm^2$. As most dosages are conventionally reported per unit area, it is convenient to report dosages determined from ferrioxalate actinometry in terms of per unit area rather than per unit volume. Dosages determined by ferrioxalate actinometry were correlated to radiometer measured energy fluxes using the experimental setup by Qualls and Johnson (1983) to establish the bioassay-dose relation. Another means for dosimetry is the bioassay method developed by Qualls and Johnson (1983) to determine UV dosage in a solution in which direct UV measurement using a radiometer was not possible.

Section 7. COMBINED PROCESSES

The beneficial effects of radiation on water, wastewater and sludges can sometimes be accomplished at a lower dose and/or more cheaply if irradiation is carried out in conjunction with another process. Several "radiation-enhancement" processes that are currently under investigation include ultra violet/peroxide, combined treatment, oxyradiation, ozone-plus-irradiation, and thermoradiation.

Oxyradiation is the combined treatment of a material with irradiation and oxygenation, where air or oxygen is usually dispersed in the material during irradiation. Ozone-plus-irradiation is the combined treatment of a material with irradiation and ozone, where oxygen containing ozone is dispersed in the aqueous material during irradiation. Thermoradiation is the heating of a material immediately prior to or following irradiation.

The use of irradiation combined with other processes involves careful consideration of the impacts on the preceding and following processes, such as combined process efficiency, cost, and ease of administration. These parameters in turn are impacted by any other processes to which the water, wastewater, or sludge is subjected before, during, and after irradiation.

Investigations of radiation-enhancement combined processes are being carried out in many national laboratories in the world, such as the Atomic Energy Agency of Japan, the Bhabha Atomic Research Centre in India, the Geiselbullach liquid sludge irradiator facility in Germany, and elsewhere.

Oxyradiation

Lessel (1985) noted the early research on oxyradiation conducted at the irradiation facility at Geiselbullach, Germany. These investigations included the effects of using air or pure oxygen, the method of application, the concentration of oxygen needed, the oxygen yield, the pathogen inactivation capability, and the influence of the oxygen concentration on the physical sludge characteristics. He noted that the synergistic effects of the oxyradiation process were significant. The resulting increased inactivation of pathogens reduced the necessary irradiation dose, while achieving the same hygienic effect. The data indicated higher capacities for similar plants and lower cost per unit volume treated.

Figure 6 illustrates the system used at Geiselbullach for aeration or oxygenation during irradiation. An injector nozzle and a double cone mixer were installed into the recirculation pipe in the pump shaft. The batch system installation was designed for a maximum throughput air or oxygen of 50 m^3/h (220 gal/m).

Several important results of the oxyradiation studies at Geiselbullach were reported: (1) the efficiency for maintaining the oxygen concentration in the sludge by continuous injection of oxygen at a low flow rate was two to three times greater than intermittent injection at a higher flow rate; (2) at a dose of 100 krad, the bacterial inactivation rate was 15 to 38 times greater at an oxygen concentration of 5 mg/l compared to irradiation without oxygenation; (3) greater concentrations of oxygen in the sludge (15 to 25 mg/l) did not increase the level of disinfection; and (4) oxyradiation treatment with a dose of 200 krad has at least the same effect as 300 krad of irradiation without oxygenation. The dose may be reduced to 150 krad or less while still obtaining the same level of pathogen inactivation.

As part of the electron-beam studies at the Miami-Dade facility, Ouiones et al (1989) concluded that in oxygenated secondary wastewater irradiated with high energy electrons, it appears that the hydroxyl radical is primarily responsible for the decomposition of organic compounds. Thus, the removal of phenol and substituted phenols should be efficient using oxyradiation.

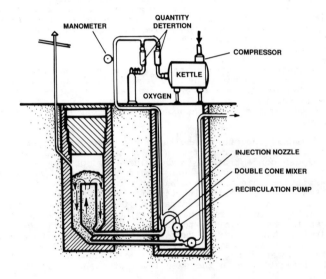

Figure 6. Schematic of system for aeration or oxygenation of wastewater sludge during irradiation.
(from Lessel, 1985)

Ozone Plus Radiation

Humic acid and fulvic acid in natural water are precursors of carcinogenic trihalomethane (THM) which is formed during chlorine disinfection in city water processing. In a study performed at the Japan Atomic Energy Institute, Takehisa, et al, (1985) evaluated the use of radiation-oxidation in the presence of ozone as a method to remove these precursors. The study concluded that, while radiation treatment of water in the presence of dissolved oxygen decomposes organics, the required dose is high. Reduction of the dose, and removal of the precursors, was achieved using ozone in combination with some conventional processes.

Unnikrishnan, et al (1986) at the Bhabha Atomic Research Centre also investigated the use of ozone and radiation on aqueous solutions. They reported a significant synergistic effect of radiation and ozone in the reduction of total organic carbon (TOC) in water containing lignin, formic acid, ethylene glycol, azo-dye, and phenol (which are difficult to degrade by activated sludge treatment). The synergistic effect was explained as the conversion of HO_2 radicals to .OH radicals which are strong oxidizing species to most of the organic compounds.

Thermoradiation

Radiation-produced charged ions and oxidizing species can affect colloidal systems and may improve the dewatering characteristics of sewage sludge. Thermoradiation may provide some synergy between the heat and radiation effects. Sandia National Laboratory (1978) reported on the biological effects of thermoradiation for raw sewage sludge samples obtained from the Albuquerque Wastewater Treatment Plant. Digested sludge samples at 4 to 5% solids were taken from the bottom of the primary digester. The samples were irradiated to doses ranging from 25 krad to 500 krad at an absorbed dose rate of approximately 60 krad/min.

The results indicated that radiation and thermoradiation appeared to be equally effective in reducing the specific filter resistance (i.e., improving the dewaterability) of raw as well as digested sludge. Radiation alone decreased the specific filter resistance of raw sludge by 0.04 to 0.49 log cycles, while the specific resistance of digested sludge was decreased by 0.12 to 0.38 log cycles. Thermoradiation of the sludges at 40 °C to 55 °C produced specific resistance decreases on the same order as radiation alone. Thermal treatment alone of the sludge samples in the range of 40 °C to 95 °C had a significant detrimental effect on specific resistance.

The report noted that the mechanism by which radiation and thermoradiation cause the specific resistance of sewage sludge to decrease is not completely understood. However, the authors suggested that free radicals caused by irradiation interact with bacterial mass, thus lysing cell walls. The cell contents, proteins and polymeric nucleic acids are dispersed. Because these released materials are similar to polyelectrolyte coagulants, they condition the sludge which results in coagulation, decreased specific resistance and increased ease of dewatering. The studies concluded that thermoradiation is generally more effective than either heat or radiation alone in the inactivation of bacteria, spores, viruses, and parasites. It was also noted that the synergistic effect of heat and radiation seen with liquid sludge was generally not seen in dewatered sludge, presumably due to uneven or inefficient heat transfer characteristics in dewatered sludge.

Section 8. APPLICATION ASSESSMENT

This review of application assessment combines a summary of information given previously on specific radiation technologies, some past and present studies, and some issues pertaining to their potential use in water, wastewater and sludge treatment. The summary compiles a list of the major irradiation installations in the world, including: the one full-scale currently operating sludge irradiator, a cobalt-60 installation to prevent biofouling of water wells, and the many research facilities which are currently examining various aspects of irradiation technology as it applies to water, wastewater, and sludge. Reference sources are given for these facilities for more detailed descriptions. The summary also includes a table of potential applications for radiation energy technology in the water/wastewater industry, best suited for each type of radiation cited: radioisotopes, electron accelerators, x-ray sources, and ultraviolet light.

Summary of Currently Operating Facilities

1. The only currently operating full-scale sludge irradiation facility in the world is the one at Geiselbullach, Germany. Its features are given below.

- liquid sludge irradiator.
- source: cobalt-60.
- in full-scale operation since 1973.
- processes 150 m^3 per day (50,000 gal/day) of sludge at 3-5% solids.
- administered dose: 300 krad.
- serves a population of 240,000 people.
- disinfected liquid sludge is spread on agricultural land for fertilizer value.
- reference: Lessel, et al (1979).

2. A system for radiation treatment of water (using cobalt-60) was installed in Germany in 760 deep drilled wells used for drinking and industrial water. The effective life of these wells is limited by build up of iron and manganese oxide hydrates on filters caused by biological and chemical action. A dose of 250-400 Gy reduces deposits by killing bacteria. A summary of its features follows.

- drinking water well filter irradiation.
- source: Cobalt-60 pencils placed in socket pipes of 2.6 cm diameter and with spacers of steel pipes interposed between the radiation sources.
- sources installed in 760 wells around the country.
- administered dose: 250-400 Gy.
- reference: Leonhardt and Wetzel (1987)

Summary of Current Research Facilities

Many research studies in waste/wastewater irradiation are underway around the world at the irradiation facilities summarized below.:

(1)   Miami-Dade County, Florida, U.S.A.

- water/wastewater/sludge irradiator.
- source: electron-beam [1.5 MeV, 75 kW beam power].
- built in 1984 and refurbished in 1988.
- processes up to 100 gal/min of wastewater containing 0.5% solids.
- administered dose: various, according to type of test being run.
- penetration in wastewater: 2-5 mm.
- research: disinfection of drinking water and wastewater, and degradation of chemicals.
- operated by the University of Miami and Florida University with support from the National Science Foundation.
- reference: Morse (1989)

(2)   Japan Atomic Energy Agency, Takasaki, Japan

- water/wastewater/sludge irradiators
- sources: Cobalt-60 & electron-beam [2 MeV, 100 kW]
- five research studies on radiation treatment for:
  1) Decomposition of refractory organics in industrial wastewater.
  2) Engineering of radiation treatment by electron beams as high power radiation
  3) Combined use of radiation and conventional treatment to improve economics.
  4) Use of synergistic effect of ozone and radiation.
  5) Radiation disinfection in place of chlorine.

- E-beam disinfects 3,000 m^3/day of effluent from a wastewater treatment plant serving 600,000 people in a research study which administers a dose of 4 kGy to disinfect and to decompose hydrocarbons (Arai and Machi, 1988)
- Cobalt-60 irradiator administers a dose of 2 Megarads to sterilize sludge cake (20% solids) for composting research (Kawakami, et al, 1981)
- E-beam disinfects dewatered sludge cake (20% solids) at a dose of 600 krad; for seeding and composting research (Hashimoto, et al, 1988)

(3) Tokyo Metropolitan Isotope Research Centre, Tokyo, Japan

- irradiation of landfill leachate.
- source: cobalt-60.
- study began in 1983.
- research: to reduce microorganisms and degrade toxic organic pollutants.
- administered dose: various.
- process: uses a combination of radiation with oxygen or nitrogen.
- reference: Sawai (1984)

(4) Czechoslovakian Nuclear Research Institute

- irradiation of drinking water
- sources: cobalt-60, cesium-137, electron-beam
- research ongoing since 1986.
- administered dose: 0.8 kGy for disinfection and 2 Gy/h for bio-fouling
- research: irradiation of drinking water to inactivate <u>Mycobacteria</u> [chlorine replacement], and to regenerate biologically clogged water wells
- reference: Vacek, et al (1986)

(5) Bhabha Atomic Research Center, India

- wastewater/sludge irradiator
- source: cobalt-60 (pilot plant of 500 kCi)
- built in 1987
- will process up to 110 m^3/day of sludge at 3% solids
- administered dose: 300 krad
- location: Gajerawadi Sewage Treatment Plant, Baroda, India.
- reference: Krishnamurthy (1984)

(6) Nordion International Inc. Canada

- irradiation of dewatered sludge (25% solids)
- source: cobalt-60
- study began in 1989
- research: to pasteurize sludge, degrade toxic organic pollutants.
- administered dose: 600 krad, 900 krad and 1,200 krad
- reference: Swinwood (1989)

(7) Austrian Research Centre, Austria

- removal of chlorinated ethylenes from drinking water by irradiation
- source: electron accelerator
- study began in 1987
- research: to reduce perchloroethylene and trichloroethylene in drinking water
- administered dose: 255 Gy (to achieve 90% reduction of perchloroethylene) and 120 Gy (for 90% reduction of trichloroethylene)
- reference: Gehringer et al (1989)

(8) Johannesburg, South Africa

- irradiation-pasteurization of chicken litter and sewage sludge to produce a pathogen-free ruminant feed
- source: 3 MeV, 120 kW electron beam
- administered dose: 10-20 kGy
- produced and sold by a private company on a full-scale basis
- reference: Scott (1986)

(9) University of Vienna, Austria

- decomposition of chlorinated phenols in water
- source: cobalt-60 research irradiator
- study conducted in 1987
- dose: up to 10 kGy
- reference: Getoff and Solar (1988)

Process Assessment by Source

A compilation of the factors which influence the choice of radiation energy source type for application in the water-wastewater-sludge treatment industry is presented in Table 2. The table highlights areas for consideration and possible further study. It is not intended to provide a definitive solution for a specific application. A few key benefits are discussed for each radiation type, as well as one or more potential problem areas or constraints.

All four sources of radiation energy have been identified in Table 2 as having potential applications in the water- wastewater -sludge treatment industry. The radioisotope sources are limited to cobalt-60 gamma sources. Electron accelerators are divided into two classes - those with energy up to 5 MeV and those above 5 MeV; x-ray sources of both types, and commercial ultraviolet light sources. Each source type has been evaluated according to three factors: (1) its state of development for industrial use; (2) main technical factors that should be taken into consideration; and (3) general economic factors which will impact on its use.

Table 2
Development Status of Systems for Irradiation
of Water, Wastewater, and Sludge

Source Type	Development Status	Technical Factors	Economic Factors
Ultraviolet	Commercially available, some pilot scale and research water treatment plants in use.	Difficult to measure actual dose delivered to control process. Sensitive to flow and mixing conditions, particulate control and optical transmission at the wavelength used.	May compete with chlorination; a 40 kW (consumption) unit can treat up to 10 Mgal (US) per day and costs (excluding the channel) about $20,000 US.
Radioactive Isotopes (Cobalt-60)	In commercial use.	Usually batch or bucket type to take advantage of the penetrating nature of the radiation. A cobalt-60, re-circulating batch, system has been in routine operation at Geiselbullach, FRG since 1973.	Relatively capital intensive and price of radioisotopes can dominate economics. Low labor, per unit and maintenance costs. Capital costs can range from $2.0 million to $7 million US. Source decay of 12.5% pa has to be allowed for if cobalt-60 is used.
Electron Accelerators 0.5 to 5 MeV	Commercial use possible; none in operation for treatment	High powers available and accelerator equipment reliability is improving. Product handling and dosimetry can be difficult due to low penetration. 1.5 MeV unit at Virginia Key, Florida being operated for R&D use. Systems available from 30 kW to 200 kW. Very thin vacuum window may corrode and damage easily.	Electricity, Operator wages and maintenance are the main operating costs. Capital costs for total system will probably range from $3 million to $7 million US.

Electron Accelerators above 5 MeV	R&D Stage	Low power equipment is in use in R&D on sewage; 10 to 20 kW units are in service in other applications; accelerators with power above 20 kW are being tested. Vacuum windows may corrode.	Capital cost of accelerator per kW higher than the 5MeV and below systems and they use more electricity. System cost at 20 kW estimated as about $4 M US rising to about $7M US for 40 to 50 kW systems.
X-ray Machines	R&D Stage.	No systems in operation applied to wastewater or sewage. Two prototype sources being developed for 150 kW & 200 kW electron, 12 kW & 16 kW x-ray output. These are both 5 MeV units.	Cost of accelerator for same capacity will be many times that of equivalent electron system because of poor conversion efficiency to X-rays. Only likely to be economic if local price for radioisotopes is very high or if the use of radioisotopes is not allowed but high penetration is needed.

Section 9. ISSUES, CONCLUSIONS and RECOMMENDATIONS

Research activity on radiation energy treatment of water, wastewater and sludges has led to an extensive literature and a vast body of experience. Some processes, such as ultraviolet energy disinfection of wastewater are relatively well-accepted. Other processes, such as x-ray, gamma-ray, and electron energy treatment are not as well known. Differences in the degree of acceptability or utilization of radiation energy treatment can be attributed to a number of factors, including technology, economics, and public perception. This Section examines some of the major issues affecting adoption of radiation treatment of water, wastewater and sludge and draws a number of conclusions about the relevance of this technology for the 1990's and beyond.

Issues

As with non-radiation treatment methods, the feasibility of radiation must be evaluated in comparison with competitive technologies. Each potential application must be assessed on its own merits for appropriate technology, cost, ease of installation and operation, its 'fit' with existing equipment and processes, and its acceptability to plant personnel and the public at large. Some of the issues which impact on the feasibility of using radiation for treatment of water, wastewater and sludge are general and apply to gamma-ray, electron, x-ray, ultraviolet radiations and their combined processes. Others affect the technology for specific types of radiation sources.

The association of radiation energy, especially ionizing radiation, with nuclear technology has had an effect on the development of radiation energy treatment technology. The cautious attitude of legislators and members of the public has resulted in slow processes for permitting, siting, and process approvals. As in other areas where the public has expressed concern, increased regulatory involvement in future development is to be expected. This generally adds costs and increases the difficulty of decision making.

Many industrial ultraviolet and one electron-beam installations are operational and have been able to resolve their regulatory and public problems. Worldwide, over 400 full-scale low-energy electron-beam and gamma-ray facilities are in use for research and industrial purposes. These installations have also been able to resolve their permitting and siting problems.

Concerns exist about the potential for reinfection by pathogenic microorganisms of disinfected sludge. Depending on the dose administered to a sludge, innocuous vegetative bacteria could be absent in the treated sludge, making nutrient competition low and increasing chances of contamination by pathogens. Irradiation treatment alone will not reduce significantly the vector attraction of sludges. In order to accomplish vector attraction reduction, irradiation can be carried out in combination with other processes.

An important issue affecting irradiation treatment using gamma sources is that of handling safety. Transport of radioactive materials on a daily basis is routine (for example, the shipment of radioisotopes to hospitals), but concerns must still be addressed. Safety of the facility containing the gamma source and of workers at the facility also concern the public, unions and plant owners. Careful, open and detailed information and discussions must be made available when considering irradiation processes.

A major technical issue for electron-beam utilization is the limited penetration with current machine configurations. For water or other media with similar density, the penetration is approximately 0.35 cm/MeV. The issue requires technical solutions for large flow rate, thin stream exposure to the beam. Successes have been achieved at the 1.5 MeV electron-beam installations in Boston and Miami for irradiating sludge with 5% solids at electron energies to about 3 MeV. For 10-MeV electron beams, facilities could be designed for stream thickness to about 10 cm, allowing greater throughput, but at higher cost per beam kW.

A major problem in the use of x-rays for radiation energy treatment is the small energy transfer efficiency. It is difficult to envisage a significant role for x-rays in water and wastewater treatment. Iverson (1987) concluded that large scale application of x-ray processing appears to require a change in economic conditions.

A major problem for effective ultraviolet disinfection is the need to maintain nearly plug flow conditions in the reactor. Ultraviolet radiation shares with the several forms of ionizing radiation the problem of not being able to provide a residual effect similar to that of chlorine. However, chlorination in combination with radiation may prove to be a more effective process than chlorine used alone. A result may be a reduced consumption of chlorine.

Conclusions

Radiation treatment can play an important role in the treatment of water, wastewater, and sludge, both directly and in combination with other processes. Results are being compiled from many operating facilities and research programs distributed worldwide. Irradiation as a process for water and wastewater treatment is under study in many technologically advanced countries. Greater application of radiation in treatment of water, wastewater, and sludge is dependent on successful demonstration of its performance compared to alternatives. Table 3 lists some of the larger irradiation facilities for Co-60 radioisotope sources and electron accelerators. For more detailed descriptions, references are cited.

With the exception of ultraviolet radiation (used primarily for clean water), applications of radiation energy in water and wastewater have been limited. Scientific study into combined processes has begun only recently. More work will be necessary before there is a reliable understanding of the advantages or disadvantages of radiation treatment in conjunction with other processes such as heat, oxygen, and other chemicals.

TABLE 3

Major High-Energy Irradiation Facilities in
Water, Wastewater, and Sludge Treatment

Location	Type	Material	Power (kW/kCi)	Cap (m³/d)	Dose (Gy)	End Use	Ref
Res. Center Austria	E-beam	drinking water			225	chem. treatment	[1]
Univ. Vienna Austria	Co-60	water with phenols	-		10	chem. treatment	[2]
Ontario Canada	Co-60	dewatered sludge (25% S)	- 12k		6K-	Pasteur	[3]
Nuc. Res. Inst Czech.	Co-60 Cs-137	drinking water	- -		800 2/hr	disinf biocide	[4]
Geiselbullach Germany	Co-60	liq. sludge (3-4% S)	650	180	3000	disinf fertilizer	[5]
760 wells Germany	Co-60	water	-	var	240-400	biocide	[6]
Bhabha ARC India	Co-60	wastewater sludge (3% S)	500	110	3000	disinf	[7]
Takasaki Japan	Co-60	sludge cake (20% S)	-		20k	compost	[8]
	2.0-MeV	sludge cake			6000	compost	[9]
	3.0-MeV	wastewater	100	3000	4000	disinf	[10]
Tokyo Japan	Co-60 +O₂ E-beam	landfill leachate	-			disinf	[11]
Inst. Wat. Res Norway	gamma E-beam	sludge + effluent				combined treatment	[12]
Johannesburg S. Africa	3-MeV E-beam	sludge	120		10-20	pasteur for sale	[13]
Florida U.S.A.	1.5-MeV E-beam	water and wastewater (0.5% S)	75	22	4000	research	[14]

REFERENCES :

[1] Gehringer, et al (1989)
[2] Getoff and Solar (1988)
[3] Swinwood (1989)
[4] Vacek, et al (1986)
[5] Lessel and Suess (1984)
[6] Leonhardt and Wetzel (1987)
[7] Krishnamuthy (1984)
[8] Kawakami, et al (1981)
[9] Hashimoto, et al (1988)
[10] Arai and Machi (1988)
[11] Sawai (1984)
[12] Berge (1989)
[13] Scott (1986)
[14] Morse (1989)

REFERENCES

Ahlstrom, S.B., Irradiation of Municipal Sludge for Agricultural Use, Radn.Phys.and Chem.J 25, 1-10 (1985).

Ahlstrom, S.B., Irradiation of Municipal Sludge for Pathogen Control: why or why not?, Radn.Phys.and Chem.J. 31, 131-138 (1988)

Arai, H. and S. Machi, Treatment of Supernatant from Sludge by Combination of Electron Beam Irradiation and Biological Treatment, JAERI, Watanuki-machi, Takasaki, Gunma, 370-12, Japan (1988).

Baker, M.N., The Quest for Pure Water, (Am. Water Works Assoc., New York, 1948).

Ballantine, D.S., Alternative High-Level Radiation Sources for Sewage and Waste-Water Treatment, in Radiation for a Clean Environment, International Atomic Energy Agency, Vienna, Austria, 309-323 (1975).

Ballantine, D.S., L.A. Miller, D.F. Bishop, and F.A.Rohrman, The Practicality of Using Atomic Radiation for Wastewater Treatment, J. Water Pollution Control Fed. 41, 445-458 (1969).

Batterman, B.W. and N.W. Ashcroft, CHESS: The New Synchrotron Radiation Facility at Cornell, Science 206, 157-161 (1979).

Bierck, B.R., S.A. Wells, and R.I. Dick, Compressible Cake Filtration: Monitoring Cake Formation and Shrinkage Using Synchrotron X-Rays, J. Water Pollution Control Fed. 60, 645-650 (1988).

Bryan, E.H., Disinfection for Minimizing Risk Associated with Management of Municipal Wastewater Treatment Plant Sludges, Proc., Fifth National Conference on Acceptable Sludge Disposal Techniques, pp. 216-222, Information Transfer, Inc., Rockville, MD. (February, 1978).

Bryan, E.H., Future Technologies of Sludge Management, Proc., 1980 Spring Seminar, Sludge Management in the Washington, D.C. Metropolitan Area, pp. 52-61, National Capital Section, Amer. Soc. of Civil Engineers (May, 1980).

Bryan, E.H., Disinfection and Indicator Organisms, Environmental Science & Technology 18, No.8 (1984).

Bryan, E.H., The National Science Foundation's Support of Research on Uses of Ionizing Radiation in Treatment of Water and Wastes, Proc., Joint ASCE-IAEA Meeting on Radiation Treatment, Washington, DC, July, 1990 (IAEA, Vienna, in press).

Buttolph, L.J., H. Haynes, and I. Malelsky, Ultraviolet Product Sanitation. General Electric Bulletin LD-14, 1953.

Carlson, D.A., R.W. Seabloom, F.B. DeWalle, T.F. Wetzler, J. Engeset, R. Butler, S. Wangsuphachart, and S. Wong, Ultraviolet Disinfection of Water for Small Water Supplies, EPA/600/2-85/092, (U.S.E.P.A., Cincinnati, OH, 1985).

Chuagin, A., N. Chuagin-Offermanns, and J. Szekely, The Effects of Ionizing Radiation on Toxic Materials in Municipal Sludges, (AECL Research Co., Manitoba, Canada, 1990)

Cleland, M.R., Chairman, Radiation Dynamics, Inc., Personal Communication (1989).

Cleland, M.R. and G.M. Pageau, Radiation Processing of Medical Devices and Food, Proc., 20th Midyear Topical Meeting of the Health Physics Society, (1987a).

Cleland, M.R. and G.M. Pageau, Comparisons of X-Ray and Gamma-Ray Sources for Industrial Irradiation Processes, Nucl.Inst. and Methods in Physics Res., B24/25, 967-972 (1987b).

Compton, D.M., S.J. Black, and W.L. Whittemore, Treating Wastewater and Sewage Sludges with Radiation: A Critical Evaluation, Nuclear News 13, 58-60 (1970).

Council on Physical Medicine Report, Acceptance of Ultraviolet Lamps for Disinfecting Purposes, JAMA. 137, 1600, (1948).

Dunn, C.G., Treatment of Water and Sewage by Ionizing Radiation, Sewage and Industrial Wastes 25, 1277-1281 (1953).

Ellis, C. and A.A. Wells, The Chemical Action of Ultraviolet Rays. Revised Ed., by F.F. Heyroth. (Reinhold Publishing Corp., New York, 1941).

EPA Design Manual, Municipal Wastewater Disinfection. EPA/625/1-86/021, (U.S.E.P.A., Cincinnati, OH, 1986).

Epstein, E. and G.B. Wilson, Composting Sewage Sludge, Proc., National Conference on Municipal Sludge Management, pp. 123-128, Information Transfer, Inc., Rockville, Md. (June, 1974).

Etzel, J.E., G.S. Born, J. Stein, T.J. Helbing, and G. Baney, Sewage Sludge Conditioning and Disinfection by Gamma Irradiation, Amer.J. of Public Health 59, 2067-2076 (1969).

Farrell, J.P., S.M. Seltzer, and J. Silverman, Bremsstrahlung Generators for Radiation Processing, Radn.Phys.and Chem.J. 22, 469-478 (1983).

Feates, F.S. and D. George, Radiation Treatment of Wastes: A Review, in Proceedings Series, Radiation for a Clean Environment, (IAEA, Vienna, 1975).

Ferger, R.E and J.T. Cowgill, Engineering Report on the Electron Beam Irradiation Project for Wastewater Sludge at the Central District Wastewater Treatment Plant of the Miami-Dade Water & Sewer Authority Department, March, 1985.

Gaudin, A.M. and M.C. Fuerstenau, On the Mechanism of Thickening, Proc., Mineral Processing Congress, London, II, 6 (1960).

Gehringer, P., et al, Removal of Chlorinated Ethylenes from Drinking Water by Radiation Treatment, Abstract for International Meeting on Radiation Processing, The Netherlands, April, 1989.

Getoff N. and S. Solar, Radiation Induced Decomposition of Chlorinated Phenols in Water, Radn.Phys.and Chem.J. 31, Nos. 1-3, (1988).

Groneman, A.F. and J.Schubert, Mechanisms of Action of Irradiation on the Conditioning of Sewage Sludge Radical Scavenging Effects, Intl.J. Applied Radn. and Isotopes 29, 301-309 (1978).

Guymont, F., Disinfection Methods, Proc., Workshop on the Health and Legal Implications of Sewage Sludge Composting, pp. 5-1 - 5-22, Energy Resources Co., Inc., Cambridge, Mass., NTIS No. PB-296566 (December, 1978).

Hare, G.E., A Review of Industrial Electric Beam Accelerators and Factors Affecting Their Use in the Irradiation of Wastewater and Sewage, Proc., Joint ASCE-IAEA Meeting on Radiation Treatment, Washington, DC, July, 1990 (IAEA, Vienna, in press).

Hashimoto, S., K. Nishimura, and S. Machi, Economic Feasibility of Irradiation--Composting Plant of Sewage Sludge, Radn.Phys. and Chem.J. 31, 109-114 (1988).

Hasit, Y. and R.I. Dick, Effects of Irradiation on Sludge Management, pp. 532-539, Proc., National Conference on Environmental Engineering, (American Society of Civil Engineers, New York, NY, 1981).

Hoather, R.C. Disinfection. J.Brit. Water Works Assoc. 42, (348) 535, (1960).

Holl, P. and H. Schneider, Disinfection of Sludge and Wastewater by Irradiation with Electrons of Low Accelerating Voltage, in Radiation for a Clean Environment, pp 123-138, (IAEA, Vienna, 1975).

Hurst, C.J., Fate of Viruses During Wastewater Sludge Treatment Processes, Crit.Rev. in Envr.Control 18, No.4, 317-343 (1989).

International Atomic Energy Agency, Proceedings Series, Radiation for a Clean Environment, (IAEA, Vienna, 1975).

International Atomic Energy Agency, Technical Document, Proc., Joint ASCE-IAEA Meeting on Radiation Treatment, Washington, DC, July, 1990, (IAEA, Vienna, in press).

Iverson, S.L., Radiation Applications Research and Facilities in AECL Research Company, Proc., 6th International Meeting on Radiation Processing, Ottawa, 1-5, (1987).

Johnson, J.D., R.G. Qualls, K.H. Aldrich, and M.P. Flynn, Ultraviolet Disinfection of Secondary Effluents, Dose Measurement and Filtration Effects. Proc., 2nd National Symposium of Municipal Wastewater Disinfection, Orlando, Florida. January 27, 1982.

Jordan, J.L. and P.R. Erickson, Dewaterability Comparison of Irradiated Vs. Nonirradiated Sludges, Rexnord Environmental Research Center (1978).

Kawakami, H., et al, Composting of Gamma-Radiation Disinfected Sewage Sludge, Radn.Phys.and Chem.J. 18, 771-777 (1981).

Kirkham, M.B. and W.J. Manning, Agricultural Value of Irradiated Municipal Wastewater Treatment Plant Sludges, Final Report to NSF, Grant No. 77-04092, NTIS No. PB 80-107865 (November, 1979).

Koller, L.R., Ultraviolet Radiation, 2nd Ed. (John Wiley & Sons, New York, 1965).

Krishnamurthy, K., Radiation Technology for Disinfection and Reuse of Sewage Sludge, Bhabha Atomic Research Centre, Trombay, Bombay - 400 085, India (1984).

Lawrence, C.A. and S.S. Block, Disinfectants, Sterilization and Preservation, (Lea & Febiger, Philadelphia, 1968).

Leonhardt J.W. and K.G. Wetzel, Radioactive Isotope & Radiation Application in the German Democratic Republic, Central Institute of Isotope & Radiation Research, Leipzig Academy of Sciences, G.D.R. Presented at the United Nations Conference for the promotion of International Co-operation in the Peaceful Uses of Nuclear Energy, Geneva, Switzerland; 23 March to 10 April, 1987.

Leboutet, H. and J. Aucouturier, Theoretical Evaluation of Induced Radioactivity in Food Products by Electron - or X-Ray Beam - Sterilization, Radn.Phys.and Chem.J. 25, 233-242 (1985).

Lemke, H.S. and A.J. Sinskey, Viruses and Ionizing Radiation in Respect to Waste-Water Treatment, pp 99-120 in Radiation for a Clean Environment (IAEA, Vienna, 1975).

Lessel, T., Optimierung des Vefahrens zur Gamma-Bestrahlung von Klarschlamm, Report No. 54, Wassergutewirkschaft der Tech. Uviversity Munchen (1985).

Lessel, T. and A. Suess, Ten-Year Experience in Operation of a Sewage Sludge Treatment Plant Using Gamma Irradiation, Radn.Phys.and Chem.J. 24, 3-16 (1984).

Lessel, T., A. Suess, and E. Englmann, Gamma-Irradiation for Disinfection and Conditioning of Sewage Sludge - The Geiselbullach Demonstration Project, Abwasserverband Ampergruppe, Eichenau, FRG (September, 1979).

Levaillant, C and Gallien, C.L., Sanitation Methods Using High Energy Electron Beams, Radn.Phys.and Chem. 14, 309-316 (1979)

Lowe, H.N., W.J. Lacy, B.F. Surkiewicz, and R.F. Yaeger, Disinfection of Microorganisms in Water, Sewage, and Sewage Sludge by Ionizing Radiation, J.Am.Water Works Assoc. 48, 1363-1372 (1956).

Luckiesh, M. and L.L. Holladay, Disinfecting Water by Means of Germicidal Lamps. General Electric Review 47, 45, (1944).

Malina, J.F., K.R. Ranganathan, B.E.D. Moore, and B.P. Sagik, Poliovirus Inactivation by Activated Sludge, in Virus Survival in Water and Wastewater Systems, Water Resources Symposium No. 7, Center for Research in Water Resources, (University of Texas, Austin, 1974).

Malkoske, G. and W. Gibson, Commitment to the Gamma Processing Industry, Beta-Gamma J., 2/90 (1990).

Markovic, V., edi, Technical and Economic Comparison of Irradiation and Conventional Methods, Final Report, Advisory Group Meeting, Dubrovnik, Yugoslavia (IAEA, Vienna, 1986).

McKeon, D., C.C. Hartwigsen, and B.D. Zak, Cesium-137 Gamma Source Consideration for Dry Sludge Irradiators, Report SAND82-0998, Sandia National Laboratories, Albuquerque, October, 1983.

McKeown, J., New Accelerator for Radiation Processing; CW-Type Linac, Proc., 16th Japan Conference of Radiation and Radioisotopes, (1983).

McKeown, J., Radiation Processing Using Electron Linacs, IEEE Trans. on Nucl.Sci., NS-32, 3292-3296 (1985).

Merrill, E.W., Water Purification by Electron Irradiation in the Presence of Polymers, Final Report, NSF Grant No. 86-12487, Massachusetts Institute of Technology, (October 1987).

Metcalf, T.G., Control of Virus Pathogens in Municipal Wastewater and Residuals by Irradiation With High Energy Electrons, Final Report, NSF Grant No. 75-14729 and 77-14454, NTIS No. PB 272347 (1977) and PB 80-104086 (1979).

Moore, B.E., B.P. Sagik, and C.A. Sorber, An Assessment of Potential Health Risks Associated with Land Disposal of Residual Sludges, Proc., Third National Conf. on Sludge Management, Disposal and Utilization, pp. 108-112 (Information Transfer, Inc., Rockville, MD, 20852, 1976).

Morris, M.E., Measurement and Calculation of Radiation Fields of the Sandia Irradiator for Dried Sewage Solids, Report SAND80-2791, Sandia National Laboratories, Albuquerque, NM, 1980.

Morse, D., Accelerating Electrons, Civil Engineering 59, No.4, 64-66, (April, 1989).

Narver, D.L., Is Sterilization of Sewage by Irradiation Economical?, Civil Engineering 27, 618 (1957).

Oda, A., Ultraviolet Disinfection of Potable Water Supplies, R.P. 2012, Division of Research, Ontario Water Resources Commission, (1969).

O'Donnell, J.H. and D.F. Sangster, Principles of Radiation Chemistry (American Elsevier, New York, 1970).

Ouiones M., et al, Removal of Phenols and Substituted Phenols from Chlorinated Wastewater Using High-Energy Electrons, Proceedings, ACS Div. Envr. Chem., Miami, Florida (1989).

Petrasek, A.C., Jr., H.W. Wolf, S.E. Esmond, and D.C. Andrews, Ultraviolet Disinfection of Municipal Wastewater Effluent. EPA 600/2-80-102, (U.S.E.P.A., Cincinnati, OH, 1980).

Qualls, K.M. and J.D. Johnson, The Role of Suspended Particles in Ultraviolet Disinfection, Applied Environ. Microbiol. 45, 872-877, (1983).

Redish, G.F., Antiseptics, Disinfectants, Fungicides and Sterilization, 2nd Ed., (Lea & Febiger, Philadelphia, 1957).

Ridenour, G.M. and E.H. Armbruster, Effect of High-Level Gamma Radiation on Disinfection of Water and Sewage, J.Am.Water Works Assoc. 48, 671 (1956).

Sandia Laboratories, The Effects of Heat, Radiation & Thermoradiation on the Filterability of Sewage Sludge, (Sandia Labs, Albuquerque, NM, January, 1978).

Sawai, T., Cleaning up Tokyo Bay, Tokyo Metropolitan Isotope Research Center, Shinagawa-ku, Tokyo, Japan, (May 1984).

Scheible, O.K., Personal Communication, 1989.

Schenck, G.O., Possibilities of Disinfecting Water by Ultraviolet Irradiation. Institut fur Strahlenchemie im Max-Planck-Institut fur Kohlenforschung, Mulheim a.d. Ruhr. Parfumerie u. Kosemtik., 1979

Schuler, R.E., Radiation Sources-EB, Radn.Phys. and Chem.J. 14, No 1/2, (1979).

Scott, P., Waste Recycling the Beta Way, (University of the Witwatersrand, Johannesburg, South Africa, January, 1986).

Seferiadis, J.P., The Effect of High Energy Electron Radiation on the Filterability and Settleability of Primary Digested Sludge, M.S. Thesis, Northeastern University, Boston, Massachusetts (1977).

Sepp, E. and P. Bao, Design Optimization of the Chlorination Process. Vol. I. Comparison of Optimized Pilot System with Existing Full-Scale Systems. EPA-600/2-81-167, (U.S.E.P.A., Cincinnati. OH, 1981).

Severin, B.F., M.T. Suidan, and R.S. Engelbrecht, Kinetic Modeling of UV Disinfection of Water, Water Res. 17, No.11, 1669-1678, (1983).

Severin, B.F., M.T. Suidan, B.E. Rittmann, and R.S. Engelbrecht, Inactivation Kinetics in a Flow-Through UV Reactor, J.Water Polln.Control Fed. 56, No. 2, 164-169, (1984).

Sinskey, A.J., D. Shah and T.J. Metcalf, Biological Effects of Irradiation with High Energy Electrons, Proc., Third National Conf. on Sludge Management, Disposal and Utilization, pp. 160-163 (Information Transfer, Inc., Rockville, MD, 20852, 1976).

Sivinski, J. and S. Ahlstrom, Design and Economic Considerations for a Cs-137 Sludge Irradiator, Radn.Phys.and Chem.J. 24, 191-201 (1984a).

Sivinski, J. and S. Ahlstrom, Summary of Cesium-137 Sludge Irradiation Activities in the United States, Radn.Phys.and Chem.J. 24, 17-27 (1984b).

Sivinski, J., S.Ahlstrom, W. McMullen, and J. Yeager, The United States DOE/EPA Sludge Irradiation Program, Water Sci. & Tech. 15, 7-24, (1983).

Spinks, J.W. and R.J. Wood, An Introduction to Radiation Chemistry, 2nd Edition (Wiley, New York, 1976).

Swinwood, J., Project Leader, Nordion International Inc., Box 13500, Kanata, Ontario, Canada, K2K 1X8 (1989).

Swinwood, J., The Canadian Commercial Demonstration Sludge Irradiator Project, Proc., Joint ASCE-IAEA Meeting on Radiation Treatment, Washington, DC, July, 1990 (IAEA, Vienna, in press).

Swinwood, J. and B. Wilson, Sewage Sludge Pasteurization by Gamma Radiation - A Canadian Demonstration Project. Radn.Phys.and Chem. J. 24, 461-464 (1984)

Takehisa, M., et al, Inhibition of Trihalomethane Formation in City Water by Radiation-Ozone Treatment & Rapid Composting of Radiation Disinfected Sewage Sludge, Radn.Phys.and Chem.J. 25, 63-71 (1985).

Thomas, J.K., Elementary Processes and Reactions in the Radiolysis of Water, in M. Barton and J.L. Magee, eds., Advances in Radiation Chemistry, Vol.1 (Wiley, New York, 1969).

Thompson, C.C. and M.R. Cleland, High-Power Dynamitron Accelerators for X-Ray Processing, Nucl.Inst. and Methods in Physics Res., B40/41, 1139-1141 (1989).

Trump, J.G. High Energy Electron Irradiation of Wastewater Residuals, Proc., Williamsburg Conference on Management of Wastewater Residuals, J.L. Smith & E.H. Bryan, eds., pp. 65-82, (Colorado State University, CO, November 1975).

Trump, J.G., E.W. Merrill, J.L. Danforth, B. deBree and K.A. Wright, Large Scale Electron Treatment of MDC-Boston Sludge - Physical and Chemical Aspects, Proc., Third National Conf. on Sludge Management, Disposal and Utilization, pp. 142-147 (Information Transfer, Inc., Rockville, MD, 20852, 1976).

Trump, J.G., High Energy Electron Irradiation of Wastewater Residuals, Final Report, NSF Grants No. 74-13016 & 77-10196, MIT, NTIS No. PB 82-101577 (1981).

Trump J.C., Energized Electrons Tackle Municipal Sludge, Amer. Scientist 69, 276-284 (1981).

Unnikrishnan, G., et al, Ozone plus Radiation on Aqueous Solution of DNA - A suggestion for Treatment of Wastewater, Radn.Phys. and Chem.J. 28, 281-282 (1986).

U. S. Environmental Protection Agency, "Proposed Rules for Disposal of Sewage Sludge", Federal Register, 54, 23, 5746-5902 (1989).

Vacek K., et al, Radiation Processing Applications in the Czechoslovak Water Treatment Technologies, Radn.Phys.and Chem.J. 28, No. 5/6, 573-580 (1986)].

Venosa, A.D. and I.J. Opatken, Ozone Disinfection - State of the Art, Proc., Pre-Conf. Workshop on Wastewater Disinfection. WPCF, (1979).

Waite, T.D., W.J. Cooper, C.N. Kuriecz, and M. Nichelson, Wastwater Treatment Utilizing Electron Beam Technology: Water Changes and Toxics Destruction, Proc., 1990 Specialty Conference on Environmental Engineering (Am.Soc.Civil Engrs., New York, NY, 1990).

Waite, T.D., Current Investigations of Water and Wastewater Treatment Utilizing High Energy Electron Beam Irradiation, Proc., Joint ASCE-IAEA Meeting on Radiation Treatment, Washington, DC, July, 1990 (IAEA, Vienna, in press).

Walden, A.E. and S.T. Powell, Sterilization of Water by Ultraviolet Rays. Proc., AWWA, p. 341, (1911).

Woodbridge, D.D., L.A. Mann, and W.R. Garrett, Application of Gamma Radiation to Sewage Treatment, Nuclear News, 60-62 (1970).

Yeager, J.G. and R.T. O'Brien, Irradiation as a Means to Minimize Public Health Risks from Sludge-Borne Pathogens, J.Water Pollution Control Fed. 55, 977-983 (1983).

Yip, R.W. and D.W. Konasewich, Ultraviolet Sterilization of Water -- Its Potential and Limitations, Water and Pollution Control 110, No. 6, 14-15, 17-18, (1972).

Yotsumoto, K., H. Sunaga, S. Tanaka, T. Kanazawa, T. Agematsu, R. Tanaka, K. Yoshida, S. Taniguchi, T. Sakamoto, and N. Tamura, High Power Bremsstrahlung X-Ray Source for Radiation Processing, Radn.Phys.and Chem.J. 31, 363-368 (1988).

INDEX

Aqueous radiation chemistry 6

Canadian Federal Government, sludge irradiation investigation 12
Cesium-137 8, 10-13, 32
Cobalt-60 8, 10-13, 30-35
Combined processes 27, 37, 39
Commercial applications 5

Dewatered sludge irradiation, x- rays 20
Dosimetry 3, 26, 27, 35

Electron accelerator 14
Electron accelerators 13
Electron-Beam applications 18
Electron-beam radiation 13
Electron-beam technology 15-18
Electron-beam treatment, 16-17; sludge disinfection, 14-15; water treatment, 15
Energy flux 27
Energy transfer efficiency, x-rays 38

Free radicals 4, 5, 30

G-value 3, 5, 7
Gamma irradiation, advantages of, 8; disadvantages of, 8; sludge treatment, 9
Gamma radiation 2, 7, 9, 11-14, 19
Gamma source 8-11, 14, 38
Geiselbullach facility 9, 10

High-energy irradiation facilities 40
High-intensity x-ray generators 19

Ionization 2, 4
Irradiation installations 30
Irradiation of dewatered sludge 20
Issues 2, 7, 30, 37

Limited penetration, electron-beams 38

Nuclear technology 37

Operating facilities 30
Oxyradiation 27, 28
Ozone plus radiation 29

Radiation chemistry of water 5
Radiation effects 4, 29
Radiation energy source types 33; development status, 35-36
Radiation treatment 1-4, 6, 8, 13, 29, 31, 37, 39
Reaction time 6
Reaction types 6
Regulatory involvement 37
Reinfection, potential for 37
Research 1, 9, 11-18, 22, 23, 27-33, 35, 37, 39
Research facilities 30, 31
Residual effect, ultraviolet radiation 38

Safety. 38
Sandia Irradiator for Dried Sewage Solids 11
Sludge Disinfection System 12

Thermoradiation 27, 29, 30

U.S. Department of Energy (DOE), applications for cesium-137 11
Ultraviolet (UV) radiation 21
Ultraviolet disinfection studies 25
UV Radiation, water disinfection 23
UV Reactors, design of 25
UV treatment 24; problems, 26

Wastewater treatment 1, 6, 10, 12, 16, 19-21, 24, 29, 32, 38, 39
Water treatment 1, 2, 6-9, 15, 18, 23
X-ray applications 19; potential use, 20
X-ray irradiation facility 19
X-rays, technology 19